Whaling in the North Atlantic
Economic and political perspectives

Proceedings of a conference held in Reykjavík on March 1st 1997

Editor
Guðrún Pétursdóttir

October 1997
Fisheries Research Institute

© 1997 by Fisheries Research Institute, University of Iceland
University of Iceland Press
Printed in Iceland

All rights reserved

ISBN 9979-54-213-6

Contents

Some key concepts 5

Whales of the North Atlantic 7
Latin and common names in English, Norwegian and Icelandic

Foreword 9
Guðrún Pétursdóttir, Director, Fisheries Research Institute

Opening Address 11
Halldór Ásgrímsson, Minister for Foreign Affairs of Iceland

Address 15
Lars Emil Johansen, Prime Minister, Greenland Home Rule

**Whale resources in the North Atlantic
and the concept of sustainability** 17
Jóhann Sigurjónsson

On the resumption of the Norwegian mink whale hunt 33
Trond Bjørndal, Jon M. Conrad & Anders Toft

Whaling and the Icelandic economy 39
Thordur Fridjonsson

The International Whaling Commission today 47
Ray Gambell

The North Atlantic Marine Mammal Commission 67
- in principle and practice
Kate Sanderson

NAMMCO, IWC and the Nordic Countries 75
Steinar Andresen

Whales, the U.S. Pelly Amendment and international trade law . . . 89
Ted McDorman

CITES and international trade in whale products 99
Jaques Berney

Whaling and international law 113
William T. Burke

**Recent developments in the IWC
aboriginal subsistence whaling category** 123
Ray Gambell

Fostering a negotiated outcome in the IWC 135
Robert Friedheim

Some key concepts

IWC
The International Whaling Commission is an international organisation that regulates whaling pursuant to the International Convention for the Regulation of Whaling (ICRW) signed in 1946. The Convention's objective is "to provide for the proper conservation of whale stocks and thus make possible the orderly development of the whaling industry."
In 1982 IWC adopted a moratorium on commercial whaling. Norway lodged an objection to the moratorium, and is therefore not bound by this decision. Iceland, which did not object to the moratorium, withdrew from the IWC in 1992. Canada withdrew in 1982. Denmark represents the Faroe Islands and Greenland. There are 39 contracting governments to the Convention.

NAMMCO
The North Atlantic Marine Mammal Commission is a regional international management body established in 1992. The members are Iceland, Greenland, Norway and the Faroe Islands; Canada, Denmark, Japan, Namibia and Russia participate as observers. So far NAMMCO has not given management advice on large whale species such as the minke, sei and fin whale. However, in 1995 sighting surveys (NASS 95) were conducted under the auspices of NAMMCO, and in March 1997 updated abundance estimates for a number of whale species (including the large whales) in the North Atlantic will be reviewed by the NAMMCO Scientific Committee.

CITES
Convention on International Trade in Endangered Species regulates international trade in whale products by listing different species on two main appendices.

Species threatened with extinction are listed on Appendix I, and international trade is not permitted. Appendix II contains species not immediately threatened with extinction, but which may become so unless trade is subjected to strict regulations. In 1983 many whale stocks and species were listed on Appendix I; one exception is the West Greenland minke whale stock.

A Norwegian proposal to downlist minke whale in 1994 was defeated. Norway and Japan have proposed to downlist several whale stocks at the CITES meeting in Harare in June 1997. Again, the proposal was defeated, albeit with a smaller margin than before.

Iceland is not a member of CITES.

PELLY

The Pelly Amendment, a US domestic law, allows the US President to impose trade sanctions against any state that diminishes the effectiveness of an international conservation program. Canada, Iceland, Japan, Norway and Russia have all been threatened with economic sanctions due to their whaling activities, but the US President has each time decided not to impose the sanctions. In December 1996 US threatened Canada with sanctions due to the taking of two bowhead whales. Canada claims that any sanction will be in violation of GATT/WTO, and if sanctions are imposed Canada will make a complaint to WTO.

UNCLOS

United Nations Convention on the Law of the Sea, Article 65: "States shall co-operate with a view to the conservation of marine mammals and inthe case of cetaceans shall in particular work through the appropriate international organizations for their conservation, management and study."

Whales of the North Atlantic

English	Latin	Norwegian	Icelandic
Atlantic white-sided dolphin	*Lagenorhynchus acutus*	hvidskjæving	leiftur
baleen whales	*Mysticeti*	bardehvaler	skíðishvalir
beluga whale	*Delphinapterus leucas*	hvithval	mjaldur
Biscayan right whale	*Eubalaena glacialis*	nordkaper	hafurkitti, Íslenskur sléttbakur
black right whale	*Eubalaena glacialis*	nordkaper	hafurkitti, Íslenskur sléttbakur
blue whale	*Balaenoptera musculus*	blåhval	steypireyður
bottlenosed whale	*Hyperoodon ampullatus*	bottlenose nebbhval	andarnefja
bowhead	*Balaena mysticetus*	Grønlandshval	norðhvalur, Grænl. sléttbakur, Grænlenskur hvalur
cachalot	*Physeter catodon* *Physeter macrocephalus*	spermasetthval spermhval	búrhvalur
Californian grey whale	*Eschrichtius robustus*	gråhval	sandlægja
common dolphin	*Delphinus delphis*	delfin	hundfiskur
common porpoise	*Phocoena phocoena*	nise	hnísa
fin whale	*Balaenoptera physalus*	finnhval	langreyður
gray whale	*Eschrichtius robustus*	gråhval	sandlægja
Greenland right whale	*Balaena mysticetus*	Grønlandshval	norðhvalur, Grænl. sléttbakur, Grænlenskur hvalur
grey whale	*Eschrichtius robustus*	gråhval	sandlægja
harbour porpoise	*Phocoena phocoena*	nise	hnísa
humpback whale	*Megaptera novaeangliae*	knølhval	hnúfubakur
killer whale	*Orcinus orca*	spekkhogger	háhyrna, háhyrningur
lesser rorqual	*Balaenoptera acutorostrata*	vågehval	hrafnreyður, hrefna
long-finned pilot whale	*Globicephala melas*, *Globicephala melaena*	grindhval	grindhvalur, marsvín
minke whale	*Balaenoptera acutorostrata*	vågehval	hrafnreyður, hrefna
narwhal	*Monodon monoceros*	narhval	náhvalur
narwhale	*Monodon monoceros*	narhval	náhvalur
North Atlantic right whale	*Eubalaena glacialis*	nordkaper	hafurkitti, Íslenskur sléttbakur
northern bottlenosed whale	*Hyperoodon ampullatus*	bottlenose, nebbhval	andarnefja
Pacific gray whale	*Eschrichtius robustus*	gråhval	sandlægja

pottwhale	*Physeter catodon,*	spermasetthval	búrhvalur
	Physeter macrocephalus	spermhval	
right whales	BALAENIDAE	retthvaler,	sléttbakar
		slettbakhvaler	
rorquals	BALAENOPTERIDAE	finnhvale	reyðarhvalir
sei whale	*Balaenoptera borealis*	seihval	sandreyður
sperm whale	*Physeter catodons*	spermasetthval,	búrhvalur
	Physeter macrocephalu	spermhval	
white whale	*Delphinapterus leucas*	hvithval	mjaldur
white-beaked dolphin	*Lagenorhynchus albirostris*	kvitnos	hnýðingur

Foreword

The latter part of the twentieth century has seen a tremendous extension of international law and regulations to cover most activities of men on the oceans. Every step forward in this law-making process has called forth disputes amongst and within nations even to the point of military intervention. None of these quarrels have managed to stir much interest outside the nations involved, with one notable exception, that of whaling.

The question of whale-hunting seems to be one of the hottest topics in the whole environmental debate everywhere and one that arouses worldwide public opinion, even in landlocked countries far away from the sea. Environmental organisations and agencies have seized upon this fact to make their campaigns against all whaling a central point in their fund-raising activities - and with great successes.

Among the western-Nordic countries there is however a near-consensus that the management of renewable natural resources must be based upon the principle of sustainability, and that this principle must also pertain to the management of marine mammals. Their point of view is that decisions on whether or not to catch whales should be built upon scientific criteria and ecological considerations.

That was also without doubt the objective of the International Convention for the Regulation of Whaling (ICRW) signed in 1946, the charter for the International Whaling Commission (IWC). It is still supposed to operate on the principle "to provide for the proper conservation of whale stocks and thus make possible the orderly development of the whaling industry." As is well known, however, the majority of the 40 member states of the IWC have in recent years refused to regulate any whaling whatsoever and instead ordered a complete moratorium on commercial whaling. Other international organisations such as CITES (The Convention on International Trade in Endangered Species) have followed in the wake of the IWC listing most whales as stocks threatened with extinction and thus prohibiting any international trade in products derived from their utilization.

In addition to this the United States Congress has passed a law, The Pelly Amendment, which puts pressure on the US President to impose trade sanctions against any state that diminishes the effectiveness of an international conservation program. Canada, Iceland, Japan, Norway and Russia have all been threatened with economic sanctions due to their whaling activities, although the President has each time decided not to impose them.

All these facts make it highly actual to bring together distinguished experts to discuss the economic and political perspectives of whaling with reference to international law and law enforcement, the practices of the relevant international organisations and the current political climate. That was the aim of the Fisheries Research Institute of the University of Iceland and The High North Alliance of Norway when they organized a conference on Whaling in the North Atlantic in Reykjavík on March 1st 1977. As can be seen in the following proceedings speakers included prominent representatives of the relevant international organizations as well as the most poignant critics of the current international regime of whaling. The conference was meant to be an informative forum for objective discussion, - not a propaganda affair for or against whaling. I hope that the proceedings may serve as a helpful guide to those who will have to make difficult decisions in these matters in the near future.

Guðrún Pétursdóttir PhD, Director
Fisheries Research Institute
University of Iceland

Opening address by Halldór Ásgrímsson, Minister for Foreign Affairs of Iceland

Ladies and gentlemen,

It is an honour to have the opportunity to address you here today. It is indeed encouraging to see the high-level of competence, in terms of speakers, on the different aspects of the issue the conference is meant to deal with.

Five years ago, I left my post as Minister of Fisheries after having held that position in the Government of Iceland for nearly 8 years. During that period, a considerable part of my time and energy was devoted to the whaling question. I participated in some of the meetings of the International Whaling Commission, made several trips to neighbouring countries to consult and negotiate the issue. In all modesty I think I can claim to have been one of the principal players in strengthening the cooperation of countries in the high north, which led to the establishment of the North Atlantic Marine Mammal Commission (NAMMCO).

All this came about because in the last 25 years or so, quite different views on the whale stocks and their utilization have evolved.

First, we have the view of those who regard the whale stocks as a renewable, exploitable resource and favour scientifically based, sustainable harvesting of this resource.

Secondly, there are those who are not willing to accept the view that whales are an exploitable resource and are convinced that whales are special animals that deserve full protection. The anti-whaling industry, which often claims this view, has in fact become a regular business in its own right, fuelled by well meaning, innocent people who donate their money to something they believe will improve our world.

The third category consists of those who in general agree that whales are an exploitable resource but think that stronger scientific evidence on the status of the stocks is still needed.

In my opinion the second group, the fanatics, is the smallest, while I think the last mentioned group, people with a more reasonable approach to the whaling question, but who are perhaps not well informed, deserves more attention. I have to admit that we have not done enough to provide information in recent years and that we still have much work ahead.

Icelanders depend for their livelihood on the sea and its resources. Marine products account for almost 80% of Iceland's total export earnings. Our waters are among the richest fishing grounds in the world, and we have done our utmost to conserve the fish stocks and increase their utilization. To do so we have applied extensive marine research programmes and the best available scientific methodology.

In our view, we should apply the same methods to marine mammals. Investigations by the Scientific Committee of the IWC, and later by the Scientific Committee of NAMMCO have indeed shown that the stocks of minke, fin and sei whales in the North Atlantic are well above harvestable levels.

Although today it is undeniable that certain whale stocks can be safely harvested, widespread and vocal calls are being made for complete protection of all whales, regardless of the state of specific stocks. These demands have been supported by various nations, particularly in the western world.

It is understandable that environmental campaigners should focus on endangered species, and it is also understandable that their arguments about whales should appeal to nations that have little acquaintance with fisheries. But bracketing all species of the same biological order together as far as utilization is concerned is clearly out of the question for communities of the high north, largely dependent on the marine resources. No one would consider, for example, enforcing a worldwide ban on fishing only because the cod population on certain banks has been endangered by overfishing. Exactly the same principle applies to marine mammals, the fact that specific whale stocks are endangered is no argument for protecting all whale stocks.

Unqualified protection of all whales and other marine mammals is also contrary to modern concepts of sustainable resource management. The 1992 United Nations Conference on the Environment and Development (UNCED) in Rio de Janeiro endorsed the basic principle that all states should commit themselves to the conservation and sustainable use of living marine resources. Nations that bear the greatest responsibility for rational utilization of marine resources cannot, therefore, accept the notion of total protection of whales.

The Rio Conference endorsed the right of states to utilize their own resources in accordance with their own environmental and development strategies. Prior to that, the United Nations Convention on the Law of the Sea acknowledged the jurisdiction of states over such utilization within their 200-mile exclusive economic zones. It also recognized marine mammals as a resource, and declared that states should cooperate with a view to the conservation of cetaceans through the appropriate international organizations for their conservation, management and study. Iceland has met its obligations in this area and will continue to do so. With respect to the International Whaling Commission, Iceland is understandably very reluctant to rejoin, as the Commission has failed to adhere to its own convention.

Ladies and Gentlemen

Clearly, international law and science as well as the modern philosophy of sustainable development are in favour of rational utilization of the resources. All responsible nations must utilize their resources with both the interests of present and future generations in mind. Coastal states with centuries of fishing experience ought to have developed the most reliable knowledge about the best way to harvest these resources.

Iceland's viewpoint has thus been, and still is, that safe harvesting of the whale stocks under active supervision and based on scientific foundation offers an economical and sustainable way of utilizing the resources of the ocean. We need to carry that message forward before it becomes history to catch whales (or even fish for that matter). It is important that countries in the north speak with one voice on this issue. Only then will we be able to establish a more tolerant view towards our value judgements and cultures, which indeed is a key to the solution of this issue.

I believe this conference will help us understand what we are dealing with when it comes to the complicated questions on the future of whaling. I am therefore pleased to announce the opening of this one-day Conference on Whaling in the North Atlantic.

Address by Lars Emil Johansen
Prime Minister, Greenland Home Rule

Ladies and Gentlemen

It is with very short notice that I have changed the programme of my official visit in Iceland in order to address the conference on an issue of importance for Greenland. Unfortunately I will not be able to participate in the panel discussion later this afternoon.

Greenland is a member of the IWC together with the Faroe Islands and Denmark. This is not the case for Iceland. The global environmental conference in Rio in 1992 adopted many good principles that we work to achieve. In one of the Rio-papers it is stated that IWC is responsible for the management of whales. Almost everybody who quotes this ends the citation at that point. This is an erroneous quotation because the sentence continues as follows: "..according to the articles in the IWC Convention of 1946."

The Rio-paper does not say anything about the responsibility of the management of whales should the IWC not follow the articles of the 1946 Convention and thereby not follow its own rules.

I believe that this issue will be taken up later on today, and I am sure that interesting answers will be given.

As you are aware of, Greenland is placed under the category of "Aboriginal Subsistence Whaling" in IWC. Some have the perception that this category is a living whaling museum. We do not! We do not export the whale meat, but use our whale resources internally in Greenland as we use our fish resources. We eat the whale meat, blubber and mattak and sell the surplus that we can not eat ourselves to our neighbours or the neighbouring town. The income is used to cover the cost of modern and efficient killing-methods. One shot amounts to about 1000 US dollars.

We still go to the IWC to negotiate our whaling-quotas. We cooperate with the IWC and have presented massive amounts of reports about our whaling to the IWC.

We presume that the IWC will treat us fair to avoid the Makah-Indian situation. In doing so, we will not be pushed to look for other parties to cooperate with in the near future.

We are many who have fought for a long time for the principle that conservation and sustainable use of seals and whales is in no way different from conservation and sustainable use of other living resources.

I believe that we can begin to see the light at the end of the tunnel.

I think that enforcing ones own emotional attitudes upon other peoples (which by the way is not legitimate according to the Rio-principles) gradually will yield for the respect for other peoples' right to live their life in harmony with nature.

I wish you a productive conference.

Whale resources in the North Atlantic and the concept of sustainability

Jóhann Sigurjónsson[1] *Marine Research Institute P.O. Box 1390, 121 Reykjavík, Iceland*

INTRODUCTION

When globally considered, the history of whaling is a rather sad story of failures in managing an important marine resource. The fact that whale stocks throughout the world were overexploited was in most cases evident due to a decline in catch rates, but the science behind such conclusions was fairly primitive until some 20 years ago. While the question of whaling has recently become more of a political or ethical question, it used to be mainly a question of the scientific assessment of whalestock status. However, although other aspects of whaling are now being widely studied, and are the main theme of this conference, the status of the resource still plays a role in the debate.

This paper firstly gives a brief mention of the concept of sustainability. Secondly, it reviews the history of the exploitation of whales in the North Atlantic Ocean. Thirdly, it reviews the state of the whale stocks and the ecology of the principal whale resources in the North Atlantic, with particular reference to Iceland.

THE CONCEPT OF SUSTAINABILITY

The concept of sustainable utilization became a key principle in the 1980´s in the important work carried out by the Brundtland Commission,

---
[1] Present address: Ministry for Foreign Affairs, Raudarárstígur 25, Reykjavik, Iceland

a United Nations appointed group of experts and politicians who were given the task to look into the future of mankind, future potentials and environmental framework. The report [1] sought to provide solutions to problems of the planet, and rather than solely presenting a pessimistic view of the situation, the concept of sustainable utilization and sustainable development of local communities was introduced and highlighted. This ideology was brought forward and became the main theme of the 1992 Rio Conference on Environment and Development (UNCED).

After the Rio Conference, the concept of sustainability has been much debated, especially after nations began to realize what commitments they took on in Rio and how to implement these commitments. Unfortunately, judging from the international discussions on whaling, it seems that the sentiments of this concept are far from being commonly recognized.

What is sustainable utilization of the resources or sustainable development of communities about? Let us consider this as a balance sheet (Table 1), where relative stability is reached between birth and growth of individual animals and populations on the one hand, and factors like diseases, competition, predation and exploitation by man on the other. When this all is in balance, we can speak of sustainable harvest or a sustainable utilization of the resource. This forms the basis for sustainable development of local communities, where the people that live near the resource are allowed and encouraged to exploit the resource and its environment, given that enough is left for future generations. So, after all, although the new concept of the late 1980's and early 1990's sounded a bit foreign to many resource users, such as the fishermen and farmers of Iceland, it fully coincides with the dominant view within these communities regarding "wise use of the resource" or "rational utilization" of the resource.

Debet	**Credit**
Diseases	Births
Competition, reduced growth	Individual growth
Predation	
Exploitation	

Table 1. The Concept of Sustainability - A Balance Sheet

In what way is the concept of sustainable development relevant for the theme of this conference, i.e. the economic and political perspectives of whaling in the North Atlantic? Firstly, it poses the question to scientists whether the history of whaling shows that sustainable utilization of whale resources is just an unrealistic ideology. Secondly, it poses the question to us whether we have become capable of using the resource wisely. Let us first look at the history of whaling.

THE HISTORY OF LARGE SCALE WHALING

The history of whaling in the North Atlantic dates back centuries [2-4]. Evidence of whaling activities in northern Norway reaches to the Stone Age. Pilotwhale drive-fisheries occurred in many places in the Northeast Atlantic several centuries ago, particularly off the Faroe Islands [5], but also around the Shetland Islands and elsewhere in the North Atlantic. Other smaller species of odontocetes have also been the targets of subsistence fisheries.

Most of these early and primitive whaling activities in the North Atlantic were probably modest in terms of the proportion of the stocks annually harvested, the single most important exception being the extinction of the grey whale (*Eschrichtius robustus*) in the North Atlantic Ocean. Some bone remains of this species have been found in countries on the European continent, and in medieval Icelandic literature the painting and description of an animal very much resembling grey whale evidently demonstrates its precense in the waters around Iceland [6].

Undeniably, the classic pattern of large whale exploitation throughout the world has been that of overhunting, where one stock was depleted after another. This took place as early as the 16th and 17th centuries, when the black right whales (*Eubalaena glacialis*) and the Greenland right or bowhead whales (*Balaena mysticetus*) were the main targets of the large European pelagic fleets operating in the North Atlantic and Arctic Oceans. In the 18th and 19th centuries the American whalers entered the scene intensively hunting sperm whales (*Physeter catodon*) and other species in smaller numbers, such as the humpback whales (*Megaptera novaeangliae*).

Unfortunately, the lessons of the past went largely unheeded during the so-called era of modern whaling, which began some 120 years ago, after

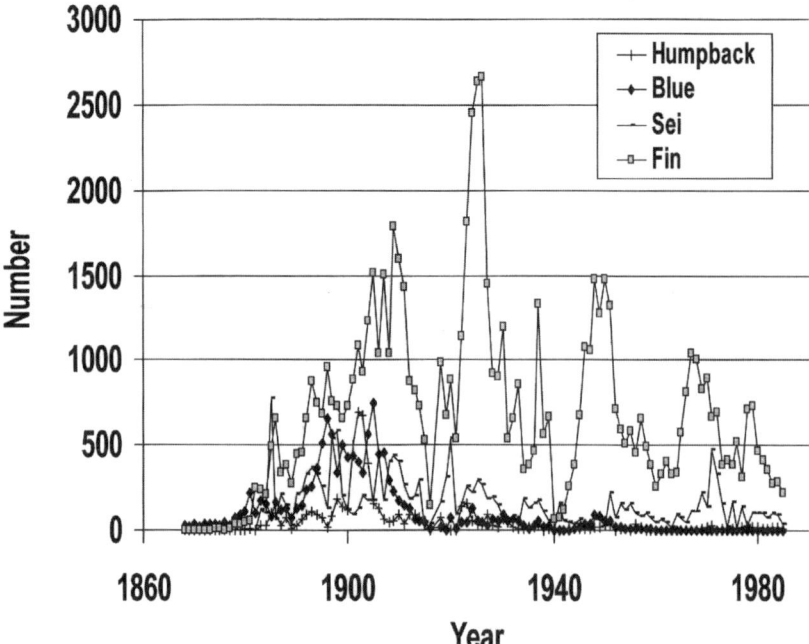

Fig. 1 Catch of blue, fin, sei and humpback whales in the North Atlantic Ocean 1868-1985, based on published information and estimated catches (based on sources in [8]).

the invention of steam ships and the explosive harpoon [7]. Now, the numerous and fastswimming rorquals, blue (*Balaenoptera musculus*), fin (*B. physalus*), sei (*B. borealis*), and humpback whales, as well as the sperm whale, became the main species of economic interest.

Fig. 1 shows the development of large whaling throughout the North Atlantic from 1868 until 1985, including Canada, Greenland, Iceland, Faroe Islands, Norway, Great Britain, Ireland and Spain (see summary and sources in [8]). Examination of the species composition indicates that some 79.000 fin whales, 12.000 blue whales, less than 10.000 humpbacks and around 16.000 sei whales were landed during commercial whaling in the North Atlantic from the late 1860s . Fin whales were most important both in terms of numbers and weight during most of the time, although the much heavier blue whales were very important during a period prior to and just after the turn of the century. Humpback whales were also taken in greatest numbers during the first period of modern whaling, while catches of fin whales increased until the mid-1920s.

Catches were reduced first as whaling activities closed down in the British Isles and further as commercial whaling ceased in the Faroe Islands in the 1960s, in Canada and N-Norway after 1972, and in Spain and Iceland in the mid-1980s. While all whale species were taken in very small numbers during the world wars and only insignificant numbers of blue and humpback whales were caught after World War I, fin and sei whales were mainly caught in the post-war period, until a temporary ban on commercial whaling was introduced after the season of 1985 [8].

MODERN WHALING IN ICELAND

Although the Icelanders had no part in the early history of whaling on the open seas, the utilization of whales in Iceland as a source of food is well documented in medieval manuscripts dating back to the 13th century, such as the well known code of law, Jónsbók. Whales were harpooned (or speared), driven ashore, or utilized when they were found naturally beached (the Icelandic word *hvalreki*, meaning literally "a stranded whale", has thus acquired its present meaning of "a godsend"). A whale often filled the desperate need for food when times were hard. The contemporary code of law defined the share of the whale belonging to the harpooner, the landlord, and even the poor.

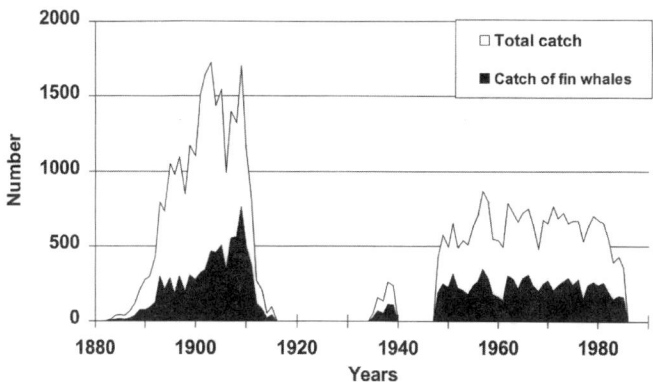

Fig. 2. Commercial whaling for large whales in Iceland (total catches and number of fin whales) from land stations 1883-1985 (based on sources in [8])

The first whaling station in Iceland, set up and owned by Norwegians in 1883 in Alftafjord, West Iceland, heralded the era of modern whaling in Iceland [7-9]. Before the turn of the century, seven land stations were located in the Western Fjords concentrating on blue, fin, and humpback whales, while sei and sperm whales were taken in smaller numbers (Fig. 2). In 1902 thirty ships landed some 1.300 whales, caught off the coast of Iceland.

As the stocks apparently became depleted, most of the stations moved to the East Coast where large whales were still in good numbers. However, after a short period of increased yield, both the total catch of blue whales and catch per vessel rapidly declined, followed by a lesser reduction in the number of fin whales caught. Aware of the clear signs of overexploitation, the Icelandic parliament (*Althingi*) proclaimed a ban on all whaling activities to begin at the close of the 1915 season. No permits were issued for whaling from land stations in Iceland until 1935, when a single station in western Iceland was allowed to operate two or three vessels. This operation ceased during World War II, and the stocks evidently recovered, at least those of fin whales.

After the war Icelandic authorities, remembering the fate of the industry at the turn of the century, issued whaling permit to only one station, located in Hvalfjordur in Southwest Iceland. The main species caught from this station was that of fin whales (on average 234 whales yearly from 1948-85), while the catch of sperm whales (82 on average p.a.) and sei whales (68 on average p.a.) was economically far less important [9]. Hunting sperm whales was prohibited in the North Atlantic by the International Whaling Commission (IWC) in 1982, following similar protective measures as taken for blue and humpback whales in the 1950s, both of which are now showing clear signs of recovery [10].

CATCH OF SMALL AND MEDIUM SIZED WHALES

Hunting of minke whales (*B. acutorostrata*), the smallest of the baleen whale species, with cold-harpoons and motor vessels did not commence until well into this century [8, 11, 36], although some primitive catches by hunters in Norway and elsewhere using nets and poisoned arrows had been practised before. At the beginning of this century the hunting was confined to small fishing boats in the nearshore waters off Norway [11] and Iceland [12, 13]. After World War II, Norwegian minke whaling expanded to the west [14] and local whaling commenced on the coasts of Canada [15] and Greenland [16, 17].

The minke whale hunt was mainly carried out by Norwegian vessels. The hunting gradually expanded towards open seas on board large and well equipped ships that were able to catch and process the whales at sea. The annual catch of the Norwegians peaked at more than 4.000 whales [18], but minke whaling was less widely distributed than whaling for other rorquals. These activities have continued to this day, although at much lower levels [15, 18-20], recently due to restrictions introduced by the International Whaling Commission (IWC). Until 1972, up to 97 minke whales were caught annually by the Canadians, and the Icelanders caught around 200 minke whales per year until 1985. Under the aboriginal-subsistence scheme of the IWC, catches in Greenland have in recent years been around 100 whales per annum.

The medium sized odontocete whales exploited in recent years [20] include the long-finned pilot whales (*Globicephala melas*) regularly hunted by driving them ashore on the Faroe Islands (500-2,500 per year), as well as 150 animals caught per year in Greenland, (in the past also off the Canadian and Norwegian coasts); a few northern bottlenose whales (*Hyperoodon ampullatus*) are taken off the Faroe Islands (thousands caught by Norwegians in the 19th and early 20th centuries) and a few killer whales (*Orcinus orca*) taken off Greenland (a few hundred were caught by the Norwegians in the post-war years).

In addition a few dolphins are caught in Greenland; over 1.000 harbour porpoises (*Phocoena phocoena*) are caught there annually as well, and an unknown number of porpoises is taken as bycatches in other fisheries throughout its range. Finally, Canada and Greenland (and Russia in smaller numbers) are annually harvesting around 500-1.300 white whales or beluga (*Delphinapterus leucas*), while Canada and Greenland each catch about 200-800 animals p.a. of the other high Arctic odontocete species, the narwhal (*Monodon monoceros*).

STATUS OF PRINCIPAL WHALE RESOURCES

When considering the small and medium sized odontocete whales, it seems that the harvesting rates [20] in recent centuries and decades have not posed serious threats to the stocks, although it must be added that detailed information is lacking for most of these species. The northern bottlenose whale apparently became seriously depleted after the turn of the century, however, recent sighting surveys indicate that some tens of thousands of animals are still left [21]. Despite the fact that pilot whales have been harvested for centuries, the best sightings estimate the stock size in the range of 780.000 whales [22, 23]. Dolphins and harbour porpoises undoubtedly number several hundreds of thousands, but scientific estimates are eithre uncertain or not available, although harvesting rates [20] are probably within the sustainable range. However, in some areas bycatches may pose problems to these stocks.

The status of the baleen whale stocks in the North Atlantic is much better known [8] and varies considerably (Table 2). As with the odontocete whales, the main source of information are data from surveys as well as data derived from catch operations [8, 22-25].

Table 2. Present Status of Stocks of North Atlantic Baleen Whales (based on sources given in [8], see also text).

SPECIES	STOCK SIZE	STATUS
Blue	1-2,000	Stock size still at low level; mainly Iceland and Gulf of St Lawrence, far less in other past grounds; increase 5% p.a. off Iceland
Fin	50,000+	Numerous, although still depleted off W Norway and UK; estimates based mainly on recent surveys, off Canada based on past tag returns
Sei	13,500+	Relatively numerous, mainly based on sightings, off Canada and US based on tag returns; depleted off N Norway
Minke	180,000+	Stock size somewhat reduced but still abundant; estimate refers to the area from Greenland to European continent; some tens of thousands may be in the NW Atlantic
Humpback	7,700+	NW Atlantic near pre-exploitation level, increasing bysome 10% p.a. in recent decades off Iceland; eastern "stock" depleted
Right/Bowhead	Few hundred	Very depleted in all Arctic areas, endangered

Large scale sighting surveys were conducted in the summers of 1987, 1989 and 1995 by the Faroe Islands, Greenland, Iceland, Norway and Spain (fig. 3 and 4). Based on information gathered there, the right and bowhead whales have not recovered after the dramatic overhunt prior to the start of modern whaling. Neither of the species numbers more than a few hundreds. They can thus be regarded as potentially endangered.

Intensive hunting of the remaining baleen whale species in the last century had a major impact on at least some of the stocks. It should be noted, however, that there are many stocks of each species, and the status differs widely from one stock to another. Apparently, the humpback whale never was in great numbers in the North Atlantic (perhaps less than 10.000)[26]. After decades of depleted status the stock has more or less fully recovered in the western and central distribution areas and off the coast of Iceland it has grown by some 10% p.a. [27, 10]. However, this species is hardly seen at all in the whaling grounds on the eastern side of the North Atlantic.

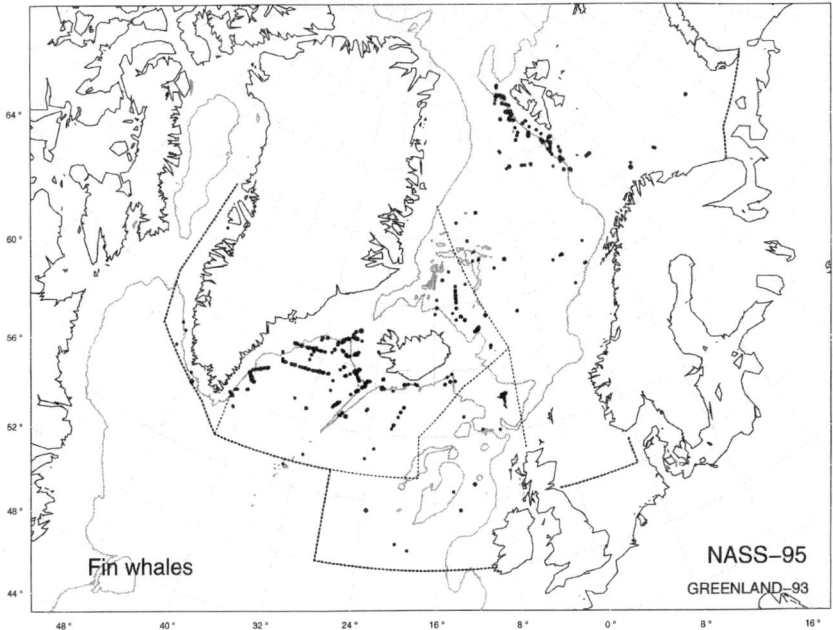

Fig.3. Distribution of sightings of fin whales during the 1995 North Atlantic Sightings Survey (NASS-95) organised by NAMMCO and carried out by Faroe Islands, Iceland and Norway (also added sightings made by Greenland in 1993, source from [23]).

It seems that prior to exploitation both sei and blue whales were in somewhat greater abundance than humpback whales, perhaps in the range of 10-15.000 whales [28]. In some areas the present number of sei whales may be of a similar order as it was prior to exploitation, while its absence from earlier grounds, such as off Northern Norway, should be noted. On the other hand, blue whales evidently are still at a low level, while showing some significant signs of recovery (5% p.a.) off the coast of Iceland [10].

Turning to the far more abundant fin whale stocks (in excess of 50.000 whales) (fig.3), which evidently were also subject to heavy taxation at the end of the last century and the first half of this century [29], the situation is generally different. This species is abundant in areas like the East Greenland/Iceland/Jan Mayen area, where some 20.000 whales occur during the feeding season [23]. However, some past whaling grounds, such as off Western Norway and the British Isles, seem still to be sparsely populated [30].

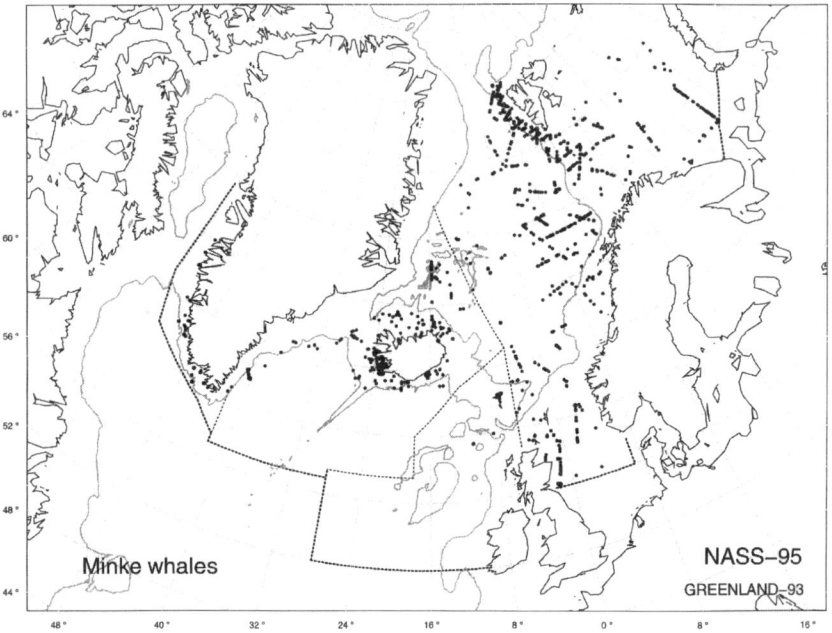

Fig.4. Distribution of sightings of minke whales during the 1995 North Atlantic Sightings Survey (NASS-95) (from [23], see further explanation in Fig 3).

Finally, the stock of minke whales (fig. 4) may originally have numbered more than 200.000 animals. It is still by far the largest baleen whale stock in terms of numbers, well in excess of 180.000 animals in the North Atlantic Ocean [23, 31]. The area between East Greenland-Iceland-Jan Mayen contains around 70.000 animals [23].

ECOLOGICAL CONSIDERATIONS

When considering biological aspects of the whaling question, apart from the status of the stocks, one additional factor most often discussed requires special mention here, i.e. the interactions between cetaceans and the commercial fisheries. In the simplest sense, such interactions can be of a direct competitive nature, where e.g. the whales eat commercially valuable fish species, or indirect, where the whales consume the food of commercially valuable fish.

Several studies have addressed the question of the consumption of whale stocks in the North Atlantic and elsewhere. One such study [32] estimated the food consumption of large and small cetaceans in Icelandic and

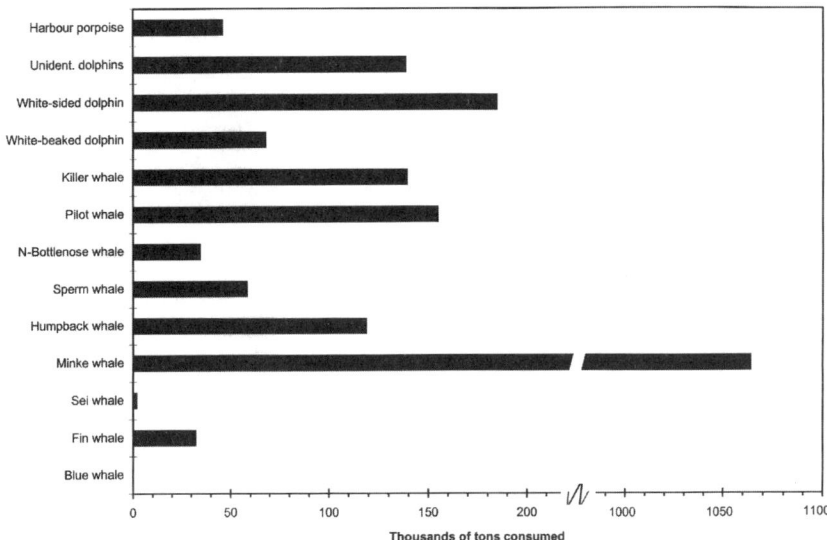

Fig. 5. Estimated total fish consumption of small and large cetaceans in Icelandic and adjacent waters (source [32]).

adjacent waters to be in the range of 6 million tons per year of animal food, whereof approximately one third or two million tons were various species of fish (fig. 5). Notwithstanding the uncertainties of such investigations, due to the various assumptions that have to be made and problems related to measuring quantities, this and other similar studies in high northern waters indicate that the amount of food consumed by whales is substantial.

However, although this figure seems large in comparison to the total annual fish landings in Iceland (1-1.5 million tons), it is far from simple to judge the potential influence of this on the commercially important fish stocks and their yields, due to the dynamic nature of interactions in the marine ecosystem. Nevertheless, the magnitude is such, that taking into account some of the major uncertainties, a model study [33] incorporating interactions of major fish and whale stocks off Iceland, shows that the long-term yield of the local cod stock (*Gadus morrhua*) will decrease significantly if the whale stocks reach their pre-exploitation levels.

CONCLUSIONS

Obviously, several of the large whale stocks were threatened by over-exploitation in past centuries, or even as late as latter half of this century. The reasons for that were a combination of several factors such as technical improvements such as new whaling barks and the introduction of steamships, build-up of too large whaling fleets in the hope of quick economic returns, and the lack of knowledge and understanding of the potential risk of catching too many whales. Finally, and perhaps not the least important reason, was the fact that the over-harvest was due to a weak organizational structure, a far too weak and poor enforcement and control mechanism, even long after the IWC was established in 1946 [34]. Unfortunately, although the IWC has in recent years shown major improvements, particularly with respect to the development of a scientific basis for whale management [34], the dominance of economic motivations in earlier times has been substituted by a fanatic protectionism with almost no tolerance for sustainable utilization of the resource. This seems to be the situation in spite of the commitments made by the world community 5 years ago in Rio de Janeiro.

Considering the dilemma that northern communities dependent on the sustainable utilization of living marine resources are facing here, it is pertinent to ask the question whether whaling is today incompatible with the sentiment of sustainable development. Has the history of overhunt deprived us the right to harvest the whale stocks? Have we got the instruments to prevent history from repeating itself?

It seems evident that the answer is yes. Firstly, investigations over the past two decades have shown that we are able to obtain information of sufficient quality to provide management advice, which if followed properly, will secure sustainable development of the stocks. Such management schemes have e.g. been developed and tested for North Atlantic minke whales by IWC scientists [34]. We also see that many of the stocks of whales occurring e.g. in the North Atlantic and discussed above, are in such numbers that they can easily stand a well regulated harvest. Indeed, many stocks are not at all endangered or threatened by extinction.

Perhaps the single most important factor for proper management of future whaling activities is the fact that the Rio Conference established the actual management aims to follow. At the same time the public awareness is such that no repetition of history will be permitted in democratic societies. Maybe it is necessary to bring the responsibility and decision making process nearer to the users of the resources, as seems to be the trend in many areas of wildlife and nature management. From the "global" IWC, albeit with less than 40 member states including landlocked states and states with almost no interests at stake or direct contribution to the research or management of whales, over to regional organizations like the North Atlantic Marine Mammal Commission (NAMMCO), which has an immediate interest in the resource [35].

Acknowledgements
I am grateful to Gísli Víkingsson at the Marine Research Institute, Reykjavík and Charlotta M. Hjaltadóttir at the Icelandic Ministry of Foreign Affairs for their help in preparing the manuscript.

REFERENCES

1. ANON. 1987. *Our Common Future.* World Commission on Environment and Development. UNEP, Oxford University Press, 365pp.
2. MACKINTOSH, N.A. 1965. *The Stocks of Whales.* The Fishing News (Books) Ltd., London.
3. BONNER, W.N. 1980. *Whales.* Blandford Mammal Series. Blandford Press, Dorset, UK, 277pp.
4. EVANS, P.G.H. 1987. *The Natural History of Whales and Dolphins.* Christopher Helm, London, 343pp.
5. BLOCH, D. 1996. Whaling in the Faroe Islands, 1584-1994: An Overview. Pp. 49-61 In: P. Holm, D.J. Starkey and J.Th. ThÛr (eds) The North Atlantic Fisheries, 1100-1976 - National Perspectives on a Common Resource. *Studia Atlantica,* 1.
6. FRASER, F.C. 1970. An early 17th century record of the Californian grey whale in Icelandic waters. *Investigations on Cetacea,* 2:13-20.
7. TÖNNESEN, J.N. and A.O. Johnsen. 1982. *The History of Modern Whaling.* University of California Press, Berkeley, 798pp.
8. SIGURJÓNSSON, J. 1995. On the life history and autecology of North Atlantic rorquals. Pp. 425-441 In: A.S. Blix, L. Wallöe and Ö. Ulltang (eds), *Whales, seals, fish and man.* Elsevier Science B.V.
9. SIGURJÓNSSON, J. 1988. *Operational factors of the Icelandic large whale fishery.* Rep. int. Whal. Commn, 38: 327-333.
10. SIGURJÓNSSON, J. and Th. Gunnlaugsson. 1990. Recent trends in abundance of blue (Balaenoptera musculus) and humpback whales (Megaptera novaeangliae) off West and Southwest Iceland, with a note on occurrence of other cetacean species. *Rep. int. Whal. Commn,* 40: 537-551.
11. JONSGÅRD, Å. 1951. Studies on the little piked whale or minke whale (*Balaenoptera acutorostrata* Lacépède). *Norsk Hvalfangst-Tid.,* 40: 209-232.
12. SÆMUNDSSON, B. 1939. Mammalia. *The Zoology of Iceland,* IV.
13. SIGURJÓNSSON, J. 1982. Icelandic minke whaling 1914-1980. *Rep. int. Whal. Commn,* 32: 287-295.
14. CHRISTENSEN, I. 1975. Preliminary report on the Norwegian fishery for small whales: Expansion of Norwegian whaling to Arctic and Northwest Atlantic waters, and Norwegian investigation of the biology of small whales. *J.Fish.Res. Bd Can.,* 32: 1083-1094.
15. MITCHELL, E. 1975. *Porpoise, Dolphin and Small Whale Fishery of the World - Status Report.* IUCN Monographs 3, 129pp.
16. LARSEN, F. and F.O. Kapel. 1983. Further biological studies of the West Greenland minke whale. *Rep. int. Whal. Commn,* 33: 329-332.
17. KAPEL, F.O. 1978. Catch of minke whales by fishing vessels in West Greenland. *Rep. int. Whal. Commn,* 28: 217-226.
18. JONSGÅRD Å.. 1977. Tables showing the catch pf small whales (including

minke whales) caught by Norwegians in the period 1938-1975 and large whales caught in different North Atlantic waters in the period 1868-1975. *Rep. int. Whal. Commn*, 27: 413-426.
19. ANON. 1988. *International Whaling Statistics*, XCV-XCVI: 1-68.
20. NAMMCO. 1993. *List of priority species*. Prepared for the NAMMCO Scientific Committee and updated at its second meeting, Reykjavík, 23-26 November, 1993, 24pp.
21. NAMMCO. 1995. Report of the Third Meeting of the Scientific Committee, Copenhagen, 31 January-3 February, 1995. NAMMCO *Annual Report*, 1995: 71-171.
22. NAMMCO. 1997a. Report of the Scientific Committee, Fourth Meeting, Tórshavn, Faroe Islands 5-9 February, 1996. NAMMCO *Annual Report*, 1996: 97-213.
23. NAMMCO. 1997b. Report of the 5th Meeting of the Scientific Committee, Tromsø, Norway, 10-14 March 1997, 105pp. (Mimeo.).
24. IWC. 1989. Report of the Scientific Committee. *Rep.int. Whal.Commn*, 39 : 33-159.
25. IWC. 1991. Report of the Scientific Committee. *Rep.int. Whal.Commn*, 41: 59-169.
26. MITCHELL, E. and R.R. Reeves. 1983. Catch history, abundance, and present status of Northwest Atlantic humpback whales. *Rep.int. Whal.Commn*, (Special issue 5): 153-212.
27. PALSBÖLL, P.J., J. Allen, M. Bérubé, P.J. Clapham, T.P. Feddersen, P.S. Hammond, R.R. Hudson, H. Jörgensen, S. Katona, A.H. Larsen, F. Larsen, J. Lien, D.K. Mattila, J. Sigurjónsson, R. Sears, T. Smith, R. Sponer, P. Stevick, and N. Öien. 1997. Genetic tagging of humpback whales. *Nature*, 388: 767-769.
28. RÖRVIK, C.J. and Å. Jonsgård. 1981. Review of balaenopterids in the North Atlantic Ocean. *FAO Fish. Ser.*, 5(3): 379-387.
29. JONSGÅRD Å. 1968. Biology of the North Atlantic fin whale Balaenoptera physalus L. - Taxonomy, distribution, migration and food. *Hvalrådets Skr.*, 49: 1-62.
30. IWC. 1992. Report of the Comprehensive Assessment Special Meeting on North Atlantic Fin Whales. *Rep. int. Whal. Commn*, 42: 595-644.
31. IWC. 1991. Report of the Scientific Committee. Annex F: Report of the sub-committee on North Atlantic minke whales. *Rep. int. Whal. Commn*, 41: 82-129.
32. SIGURJÓNSSON, J. and G. Víkingsson. 1997. Seasonal abundance of and estimated food consumption by cetaceans in Icelandic and adjacent waters. *J. Northwest. Fish. Sci.* (in press).
33. STEFÁNSSON, G., J. Sigurjónsson and G. Víkingsson. 1997. On dynamic interactions between some fish resources and cetaceans off Iceland based on simulation model. *J. Northwest. Fish. Sci.* (submitted).
34. GAMBLE, R. 1997. The International Whaling Commission today. Published in this volume.
35. SANDERSON, K. 1997. North Atlantic Marine Mammal Commission. Published in this volume.

36. ÖIEN, N. and I. Christensen. 1995. *Balaenoptera acutorostrata* Lacépède, 1804 - Zwergwal. Pp. In: D. robineau, R. Duguy, and M. Klima (eds), *Hanbuch der Saugetiere Europas*. Aula-Verlag Wiesbaden.

On the resumption of the Norwegian minke whale hunt[*]

Trond Bjørndal *Professor of Fisheries Economics and Director of the Centre for Fisheries Economics, Norwegian School of Economics and Business Administration*
Jon M. Conrad *Professor of Resource Eonomics, Cornell University*
Anders Toft *former Researcher at the Centre for Fisheries Economics,*

THE BACKGROUND

In June of 1993, Norway resumed the commercial hunt for minke whale, from a stock in the Northeast Atlantic which migrates along the coast of Norway into the Barents Sea. This was not an easy decision for the government of Norway. The International Whaling Commission (IWC) had met in Reykjavik, Iceland in 1991, in Glasgow, Scotland in 1992, and in Kyoto, Japan in 1993, and on all three occasions voted to continue the moratorium on commercial whaling, which it had introduced in 1986. The stated intention of Norway to resume whaling in 1993 brought protests from various environmental and animal rights groups, threats of economic sanctions from nonwhaling countries and threats to boycott the Winter Olympic Games, scheduled in Lillehammer, for February of 1994. What could have caused the government of Norway to risk its status as a world environmental leader, to invite the possible imposition of trade sanctions and to chance the tarnishing of the Olympic Games? Obviously, the citizens and leaders of Norway must have had strong feelings about the legitimacy and importance of whaling. For Norway, whaling is considered a management issue rather than an environmental issue.

---

[*] The authors thank Wenche Sterkeby and Margrethe Slinde for research assistance, and Nils Øien for comments.

At the Reykjavik meeting in 1991, Norway, Iceland and Japan all presented evidence that certain unit stocks might be classified as large enough to support a managed harvesting, under the prevailing IWC classification scheme. In the southern hemisphere, the stock of minke whales was estimated in excess of 600.000 animals [Woods Hole Oceanographic Institution, 1989]. In the North Atlantic, the IWC recognized four unit stocks: the Canadian east coast stock, the west Greenland stock, the central North Atlantic stock, and the Northeast Atlantic stock. The adult population in the Northeast Atlantic stock had been estimated by Conrad and Bjørndal (1991) at just under 60.000 animals.

Norway requested a resumption of commercial whaling in 1992. Japan proposed a commercial hunt for the minke whale in the southern hemisphere, and Iceland sought to resume the commercial harvest of fin whales within its territorial waters. The IWC rejected all three proposals on the grounds that its scientific committee was developing a new stock assessment procedure and that the decision to resume commercial whaling should await additional information and reclassification.

Iceland withdrew from the IWC in December 1991, feeling that the Commission had been captured by preservationists who would continue the moratorium, regardless of the mounting scientific evidence that certain stocks could be safely harvested. When the moratorium was extended again at the Glasgow meetings, Norway announced its intention to resume commercial whaling in 1993.

The decision of the Norwegian government to resume whaling was premised on two assumptions: (1) that the stock of minke whales was abundant, and (2) that they could be harvested on a sustainable basis, without risk of extinction. Given these two assumptions, the government decided to design a hunt that would be modest in the overall level of harvest, and equitable in its distribution.

An international group of scientists appointed by the scientific committee of the International Whaling Commission (IWC) in May 1996 reached a consensus on an estimate on a total stock in 1995 of 112.000 minke whales in the Northeast Atlantic, with confidence limits of 90.000 to 135.000 (High North Alliance). However this estimate has not yet been officially approved by the scientific committee of the IWC.

In the remainder of this paper we review simulation results that show, indeed, that the actual harvests over the 1993-96 period were modest, and that the stock would continue to increase through the year 2010 if harvests were fixed at 600 whales per year for 1997 through 2009. We then examine economic aspects of the 1993-1996 hunts.

STOCK DEVELOPMENT

The results of a simulation by Bjørndal and Conrad (1997) are shown in Figure 1. From an adult population of 82.100 in 1938, the comparatively high harvests in the post war period caused the population to decline to 51.900 adults in 1973. From 1970 to 1983, however, harvests ranged tightly around an average of 1.720 whales, and the population stabilized between 51.000 and 53.000 adults. After 1983, harvests drop sharply and the stock increases from 52.500 in 1983 to 63.000 in 1995.

Based on this simulation Bjørndal and Conrad (1997) conclude that there were *three stages* in the hunt for the minke whale in the Northeast Atlantic. The first, from 1938 to about 1970 was a mining or depletion stage, where the stock of adults was driven downward from its pristine population of about 82.000 to about 52.000. From 1970 to 1983, harvests, averaging about 1.800 whales per annum, stabilized the population between 51.000 and 53.000 adults. Both these stages might be consistent

Fig. 1. The Base-Case Simulation of Stock Size from 1931 to 2009.

with an open access model of the industry. Since 1983 the reduced commercial harvests, and the relatively insignificant scientific harvests during the moratorium, have allowed the stock of adults to increase to 63.000, by the beginning of 1995.

As part of their simulation, Bjørndal and Conrad (1997) took their base-case and explored the consequences of harvesting 600 adults in each year, beginning in 1997 and running through the season in 2009, in addition to the actual harvest for earlier years. This harvest seems consistent with the expected quotas for the near future. The simulation is shown in Figure 1.

From a stock of 63.000 in 1995, the adult population exhibits monotonic growth reaching 70.700 by the year 2010. Thus, a harvest of 600 adults per year would appear *in no way* to threaten the continued recovery of the minke whale population.

EXTERNAL EFFECTS OF THE RESUMED HUNT

The resumption of Norwegian whaling led to international protests from environmental and animal rights groups such as Greenpeace and Sea Shepherd, threats of economic sanctions from the United States and boycotts of Norwegian industry and the Winter Olympic Games in Lillehammer, in February of 1994. The actions were in the form of consumer boycotts, as welll as the threat of economic sanctions from the United States. The boycotts were partly directed at consumers, attempting to influence them to stop buying Norwegian products. Actions were also directed at firms importing, processing or distributing Norwegian goods. Furthermore, tourist travels to Norway were discouraged.

The effectiveness of boycotts depends on a number of factors. These include the form of the boycott (consumer boycott vs. government sanctions), the duration of the boycott, attention from mass media and resources "invested" in the boycotts. Although Norwegian whaling commanded substantial international attention in 1993, the interest for the issue gradually faded as did any effect of the boycotts.

The composition of exports in terms of e.g. bulk goods or branded goods is also important. The fact that a very large share of Norwegian exports consists of raw materials (e.g. oil, gas and fish) and semi-finished products, made Norway less vulnerable to boycotts.

To what extent rival firms and rival products as well are boycotted is important for the market share the producer may loose. Lack of a boycott of rival firms can be an important reason for large costs due to a boycott, but there are exceptions. One important exception is fish and fish products. The supply of unprocessed fish, which is the basis for the processing industry, is to a large extent given, at least in the short run. The quantity produced is therefore to a great extent determined by nature and government regulations. If for instance a boycott in Germany caused the competing nations to increase their market share at the expense of Norway, then they have to reduce their market share in other markets, and as a result, new markets would open for Norwegian fish. In this situation the costs of a boycott are limited to the costs connected with change of market.

Finally, due to increased publicity, boycotts may have a positive effect, e.g. for tourism. Actually, tourism from Germany and the United Kingdom, the two countries where boycotts were most prevalent, showed a positive development. However, it could also be the case that negative effects due to whaling were counterbalanced by a positive effect due to the Olympic Games at Lillehammer.

On the basis of complaints submitted to the Norwegian Ministry of Foreign Affairs, Bjørndal and Toft (1994) investigated 41 firms that might have suffered losses due to boycotts. Fourteen instances of likely losses were found; seven related to sales to Germany, four to the United Kingdom and the remaining three to the United states. Total losses - reduced profits and increased costs attributable to the boycotts - for the 14 cases were estimated to be in the range 6.1-9.8 million NOK. The authors found that losses were mainly of a short run nature. Evidence of long run losses was not found.

Under the Pelly Amendment to the Fishermen's Protective Act, the United States may impose trade sanctions on countries that conduct fishing operations that threaten an international managed fishery. In such cases, the Secretary of Commerce is to "certify" such a fact to the President. Upon receiving a letter of certification, the President has the discretion of directing the secretary of the Treasury to prohibit the import of fish or wildlife from the offending country. Within 60 days of receiving certification the President must report to Congress on the action taken or the reasons for inactive.

On August 5, 1993, pursuant to the Pelly Amendment, the U.S. Secretary of Commerce certified that Norway's resumption of whaling undermined the effectiveness of the IWC's international conservation regime. Obviously, this caused great concern in Norway, in particular with regard to fish exports to the United states. However, President Clinton, while being opposed to whaling and expressing that U.S. trade restrictions would be justified, decided that "our objective can best be achieved by delaying the implementation of sanctions until we have exhausted all good faith efforts to persuade Norway to follow agreed conservation efforts." (Ek and Buck, 1996). Notwithstanding this decision, both the House of Representatives and the Senate agreed on condemning commercial whaling.

THE FUTURE

In our assessment, the "first steps" in the resumption of commercial whaling were conservative and responsibly taken. We suspect that the next steps will be responsible as well, ensuring a sustainable harvest. The effects of consumer boycotts were short lived. Gradually, the international community has accepted Norwegian whaling. Further increases in harvest quotas will be based on market conditions.

REFERENCES

CONRAD, J.M., and T. Bjørndal. 1991. *Economics and the Resumption of Commercial Whaling*. Norwegian School of Economics and Business Administration, Centre for Fisheries Economics, Helleveien 30. N-5035, Bergen-Sandviken, Norway, 36 pp.
CONRAD, J.M., and T. Bjørndal, 1997. *A report on The Norwegian Minke Whale Hunt*. Norwegian School of Economics and Business Administration, Centre for Fisheries Economics, Helleveien 30. N-5035, Bergen-Sandviken, Norway, 32 pp.
BJØRNDAL, T. and A. Toft, 1994. *Økonomiske verknader av boikottasksjonar mot norsk næringsliv grunna norsk kvalfangst* (Economic Effects of Boycotts of Norwegian Industry due to Norwegian Whaling). Mimeo, Bergen, February 28, 1994. (In Norwegian)
EK, C. and E.H. Buck. 1996. *Norwegian Commercial Whaling: Issues for Congress*. Congressional Research Service - Library of Congress, CRS Report for Congress, December 31, 1996.

Whaling and the Icelandic economy

Thordur Fridjonsson *Managing Director, National Economic Institute, Iceland*

INTRODUCTION

It gives me great pleasure to have this opportunity to address this distinguished gathering today on the future of whaling in the North Atlantic. This is an issue of considerable importance to Iceland, both political and economic.

I will mostly confine my remarks to the economic aspects of this issue but because of the great uncertainties involved it is next to impossible to avoid making political judgements. The central economic question we face in this context is the following: Is it worthwhile for us Icelanders to begin again to harvest whale on a commercial basis in light of the potential costs we could suffer? My primary task is thus to identify and compare the potential benefits and costs of resumed Icelandic whaling.

On the benefit side, I will first review the potential benefits from the resumption of whaling in terms of both export earnings and jobs. Next I will attempt to assess the impact a continued moratorium on whaling could have on the yield from fish stocks in Icelandic waters, especially the cod stock. On the cost side, I will first look at the potential impact of the resumption of whaling on Icelandic tourism. Then I will review the possible impact on our exports of fisheries products. Finally, I will offer some preliminary conclusions.

DIRECT ECONOMIC BENEFITS FROM WHALING

In considering the potential economic benefits from the resumption of commercial whaling, we must first establish whether there are likely to be adequate markets for whale products. Here there is little to go on. Despite the fondness of Icelanders for certain whale products, the domestic market is for all intents and purposes insignificant. The question is, therefore, whether Japan would, as it did in the past, provide a market for exported whale products. I do not know the answer to this question but parties with an interest in the resumption of commercial whaling around Iceland maintain that the market is still there. In any case, I will, for illustrative purposes, assume that it would prove possible to sell whale products at similar prices as in the past, if we were to resume whaling operations.

We can get an idea about how extensive whaling operations could become by looking at records for the years prior to the imposition of the ban on commercial whaling in 1985. We have the most detailed information for 1980-85. In this period, whale products on average accounted for less than 2 per cent of marine exports, which in turn accounted for more than 70 per cent of merchandise exports on average. The share of whale exports in marine exports peaked at 2.5 per cent in 1983. At current prices, the annual value of whale exports in the early 1980s fluctuated between 1.3-2 billion kronur. Also, in 1980-85, the number of jobs associated with whaling, both in harvesting and processing, averaged close to 100.

Were a decision to be taken to resume commercial whaling on a sustainable basis, it seems reasonable to assume, in the absence of any clear evidence to the contrary, that the whale harvest could reach the levels of the early 1980s and that export earnings could also reach similar levels, on the important assumption that market demand for whale products still exists. This means that, as a first estimate, we could expect additional export earnings of 1.5 billion kronur annually in the long run and the same number of jobs associated with whaling as before, around 100, as the hunting and processing technologies would be the same.

Jumping ahead of myself a bit, let me say that these are indeed not large numbers. For example, in 1997, we expect marine exports to amount to close to 100 billion kronur and the labour force to number more than 130 thousand persons.

INDIRECT ECONOMIC BENEFITS FROM WHALING

In addition to the direct benefits from additional export earnings and jobs, indirect benefits could arise from the interaction of commercial whaling and the size of whale stocks and the associated impact on the size of fish stocks. Around a dozen species of whales are regarded as common around Iceland, meaning that they number at least in the several thousands. The food consumption of this large number of big mammals is bound to be very substantial and could grow larger still if the growth of whale stocks is left unchecked by commercial whaling.

Some whales feed off krill while others eat capelin and/or cod. The interactions between the various species of whales and commercially important fish stocks are exceedingly complex, and I profess no expertise in this area. However, scientists at the Marine Research Institute have reported that, based on simulations using multi-species models, unrestrained whale stocks around Iceland could lead to up to 10 per cent decline in the sustainable yield from the cod stock.

This result is of course highly uncertain. It can, nevertheless, be taken as an indication of the magnitude of the indirect benefits that could over time accrue from the resumption of commercial whaling. The annual sustainable yield of the cod stock has been estimated as 350 thousand tons. The unchecked growth of whale stocks could, therefore, cost us a cod catch of 35 thousand tons per year, which in terms of export earnings is equivalent to some 4 billion kronur.

One needs to interpret this figure carefully. In the case of resumed commercial whaling, these indirect benefits would not accrue immediately but only once a sustainable equilibrium had been established between the cod catch and the whale harvest. In the meantime, we can infer that the potential indirect benefits from a resumption of whaling are likely to outweigh the direct benefits. On the other hand, the total benefits are unlikely to exceed more than 5 billion kronur per year, which is less than 3 per cent of the annual earnings from exports of goods and services and around 1 per cent of GDP.

These speculations do not include any potential loss due to "giving in" to pressure groups that oppose any utilisation of whale resources. It is argued that such a retreat puts responsible fishing in general at risk. In other words, a kind of a defence line for the right to harvest the resources of the sea on a sustainable basis would be damaged. This is a highly strategic issue which I believe is very sensitive to how the policy of whaling will be implemented. I am, however, not convinced that an immediate resumption of whaling is essential in this respect.

TOURISM AND WHALING

Let me now turn to the potential costs associated with the resumption of commercial whaling. Essentially these costs arise in the form of lost markets, either for tourism or fish products. I'll discuss tourism first.

Legitimate concerns have been raised about the impact the resumption of commercial whaling would have on Iceland's image abroad. However, quantifying this impact is exceedingly difficult, but one can perhaps get a feel for the order of magnitude by looking at the economic significance of the tourism sector for the national economy.

The tourism sector catering to foreign visitors has indeed become a very important part of the Icelandic economy. Around 200 thousand foreigners visit Iceland annually and tourism generates close to 20 billion kronur in foreign exchange earnings each year - equivalent to around a fifth of the export earnings of the fisheries. Furthermore, thousands of persons are engaged in transporting and otherwise serving the foreign visitors.

What impact would the resumption of whaling have on the number of foreign tourists coming to Iceland? There is no hard and fast answer to this question. It is notable, however, that almost two-thirds of the visitors come from non-Nordic European countries and North America. Three countries, Germany and the United Kingdom in Europe, and the United States account for almost one half of all the foreign visitors, and it so happens that opposition to whaling is particularly virulent in these countries. I certainly don't want to imply that the resumption of whaling would cause a significant drop in the number of foreign visitors to Iceland. Evidence from Norway, for example, does not substantiate such claims. However, given that there is no change that our whaling operations would go unnoticed across the world and given that visitors to Iceland tend to be socially conscious and nature friendly, it could have important consequences for our tourism trade, at least in the short run.

Let me just to give a numerical example: a 5 per cent decline in tourism earnings would cost the economy around one billion kronur. This figure is, of course, highly uncertain and, thus, only serves the purpose of putting this issue in context of potential direct gains from whaling. Somewhat larger decline, or 7-8 per cent, would be of the same order of magnitude as the direct gains.

I have so far not mentioned the fledgling branch of tourism labelled whale-watching. It is only because in the broad setting of the overall economy it is not particularly important although it may be of some local significance. There is also no hard evidence indicating that whaling and watching are mutually exclusive.

MARINE EXPORTS AND WHALING

There is no doubt that when it comes to the potential costs associated with the resumption of whaling our marine exports are at greatest risk. The fisheries remain the backbone of our exports and any setback to them has and will continue to have repercussions throughout the economy. Marine exports account for close to 75 per cent of all merchandise exports and more than one half of all export earnings.

Again it is important to note that some of our most important markets for fish products are in countries where opposition to whaling is the strongest such as the United Kingdom, the United States and Germany. Of our largest markets, only the Japanese would seem to be immune to the whaling issue. Take the United States for example. Survey responses indicate that the resumption of whaling would negatively influence the purchases of Icelandic fish products by a sizeable fraction of consumer. Although such surveys are admittingly difficult to interpret, they should be taken seriously in light of the interests at stake.

American environmental groups, perhaps especially those that have adopted the whale as their rallying call, are adept at organising boycotts, and not only by households but also by larger restaurant firms and by school districts and hospitals. Such threats are neither to be ignored nor should it cause one to retreat immediately. Boycotts are difficult to organise and, generally, such actions don't have lasting effects. The Norwegian experience is, for example, not very decisive in this respect. However, one must keep in mind that Iceland is both more vulnerable than Norway and more is at stake here. The Icelandic sales organisations maintain that they suffered the loss of contracts as a result of boycotts organised to protest our scientific whaling in the late 1980s. In addition to organised boycotts, our fish exports could be subject to official trade sanctions from, for example, the United States government, were we to resume commercial whaling. Serious sanctions of this form are, however, rather unlikely.

The danger to our best markets for fish exports from the resumption of whaling is twofold. First, the markets might shrink as consumers boycotted products clearly associated with Iceland. Second, our access to these markets could be limited by trade sanctions. I have no doubt that we would over time be able to offset any loss of markets and sell our entire output of fish products elsewhere, but I am less convinced that we would continue to obtain the same prices. The cost would, of course, be the same whether it came through reduced output or lower prices.

Again putting a number on the potential cost is subject to great uncertainties. An average decline in the prices for fish products of 5 per cent would, for example, cost the economy 5 billion kronur annually; larger price drops would lead to correspondingly larger costs.

CONCLUSION

Let me summarise briefly and then offer some preliminary conclusions based on my assessment of the economic issues involved.

In a perfect world, the potential benefits of resuming commercial whaling would be considerable, perhaps of the order of 5 billion kronur annually in terms of increased export earnings. However, the lion's share of these benefits is indirect in the sense that it would flow from a potentially larger sustainable cod catch rather than the commercial importance of a resurrected whaling industry.

On the other hand, opposition, sometimes virulent, to whaling among most of our trading partners, other than Japan, implies that the resumption of commercial whaling entails considerable risk to economically important sectors, such as tourism and the fisheries. It is not hard to imagine scenarios under which the losses to these sectors, and hence the cost to the economy as a whole, would far exceed the potential benefits that could accrue from resumed whaling.

In light of these circumstances, it is imperative that we move cautiously and properly weigh the potential benefits and costs. In order to lessen the costs, we need to co-operate with similarly inclined countries in continued efforts to convince the public on both sides of the Atlantic that, like other resources of the sea, whales should be harvested in a sustained manner that takes proper account of environmental considerations.

It is true that opposition to whaling in both the United States and Europe has probably not softened significantly over the last decade. That does not, however, mean that this will always be so. The more thoroughly the abundance of whales in the oceans is demonstrated the better chance we have of gaining supporters for our cause. Continued research on the condition of the whale stocks in the North Atlantic is essential. Such efforts in recent years has changed the nature of the discussion, although opposition to whaling is still very strong. Now, for example, arguments based on the risk of extinction and the intelligence of whales do not appeal to people as before. Hence, balanced arguments based on science are gaining ground among informed people. This indicates that time is on our side. A well prepared policy of whaling with strategic implementation, where timing is essential, might thus soon be possible without great risk.

We stand no chance of convincing the more fanatical elements of the whale rights movement that harvesting whales can be justified under any circumstances. However, the largest number of people that currently oppose whaling are not fanatics but perhaps not too well informed about the true condition of whale stocks. Among this group are potential allies whom we need to convert to our cause by emphasising the scientific basis of our sustainable approach to utilising and conserving marine resources.

The International Whaling Commission today

Ray Gambell *Secretary to the International Whaling Commission, The Red House, 135 Station Road, Histon Cambridge, UK CB4 4NP*

BACKGROUND AND OBJECTIVES

The 1946 International Convention for the Regulation of Whaling was drawn up as an agreement to control the catching operations of a particular fishery which throughout its long history "has seen over-fishing of one area after another and of one species of whale after another to such a degree that it is essential to protect all species of whales from further over-fishing." In particular the Governments involved recognised "the interest of the nations of the world in safeguarding for future generations the great natural resources represented by the whale stocks." The Convention sought to "establish a system of international regulation for the whale fisheries to ensure proper and effective conservation and development of whale stocks on the basis of the principles embodied in the provisions [of a whaling Agreement and Protocols signed in 1937, 1938 and 1945] to provide for the proper conservation of the whale stocks and thus make possible the orderly development of the whaling industry" *(quotations from the Preamble to the Convention).* The Contracting Governments agreed to establish an International Whaling Commission to implement these ideals, which is now interpreting the Convention much more as a conservation instrument in tune with the current environmental ethic, although the restrictions imposed in implementing this approach are not universally agreed or accepted by some of the communities most affected (Gambell, 1995, p.705).

The authors of the 1946 Convention were in a number of ways quite forward-looking in the language and concepts they embraced. The current ideals of wise use and sustainable development are already to be found in their text, although because of the reigning political and economic climates and the lack of fully credible scientific understanding, it was many years before they were applied and implemented in anything more than a rather nominal fashion.

Because it was recognised that the regulations governing the conduct of the whaling operations would need to be changed from time to time the practical rules determining such elements as the whaling areas, seasons, species, numbers and sizes of whales permitted to be caught are set out in the Schedule to the Convention, a document which can be amended by the Commission, normally at an Annual Meeting. Such amendments require a three-quarters majority of those Contracting Governments voting to be adopted, and there is an objection procedure for governments which may feel that their sovereign interests are unduly affected by such changes.

THE SPECIES COVERED

The main activity of the IWC during the early years of its existence was setting catch limits for the stocks of the great whales hunted in the Antarctic. This was done initially in terms of Blue Whale Units - the approximate oil yield equivalent of the different species whereby 1 blue whale was equal to 2 fin whales, $2\ 1/2$ humpback or 6 sei whales. Catch limits by species were gradually introduced in other whaling areas world-wide, and in 1972 by species in the Antarctic.

The 1946 Convention does not define a 'whale', although a table of names in English, French, Dutch, Russian, Norwegian, Danish, Swedish and Spanish corresponding to the Latin names of the larger species of whales drawn up by a Committee on Use of Scientific Names was attached to the Final Act of the Convention with the recommendation "that the chart of Nomenclature of Whales annexed to this Final Act be accepted as a guide by the Governments represented at the Conference" (*Recommendation IV of the International Whaling Conference*). These species are the right (including bowhead) whales, pigmy right, humpback, blue, fin, sei, Bryde's, minke, sperm, and the Arctic and Antarctic bottlenose whales.

Some governments, including Brazil, Chile, Denmark, Japan, Norway, and St Vincent and The Grenadines, take the view that the IWC has the legal competence to regulate catches only of these named great whales, and regard this list as definitive and limiting. Other governments, including Austria, Germany, Ireland, Netherlands, New Zealand, Oman, Switzerland, UK and USA, believe that the list was developed as an aid to avoid confusion during the negotiations in 1946 and that all cetaceans, including the smaller dolphins and porpoises, also fall within IWC jurisdiction.

Despite these differing views amongst the member countries over the question of legal competence, the IWC does recognise the need for international co-operation to conserve stocks of small cetaceans. An example of this occurred in response to an unprecedented catch of 916 killer whales by the Soviet Antarctic whaling fleets in the 1979/80 season. Although the Scientific Committee recognised that there is considerable uncertainty over the stock identity and abundance of this species, it recommended a zero catch limit. The Commission adopted this proposal by adding wording to the ban on factory ship whaling then in force to clarify that "This moratorium applies to sperm whales, killer whales and baleen whales, excluding minke whales" (IWC, 1981, p.21).

More recently the Commission has agreed that the Scientific Committee should draw together all available relevant information on the present status of those stocks of small cetaceans which are subjected to significant directed or incidental takes and on the impact of those takes on the stocks, and to provide such scientific advice as may be warranted (IWC, 1991, p.48). As part of this programme the Scientific Committee has investigated many species and stocks, and has carried out major reviews of catches of small cetaceans. The IWC has also encouraged countries to seek scientific advice on small cetaceans from the IWC, and invited IWC member nations to provide technical or financial assistance to countries with threatened small cetaceans stocks (IWC, 1993, pp.36-7, 51).

Because of the obvious need to overcome the legal problems which are preventing effective global management of small cetaceans, the IWC established a Working Group which met before the 1993 Annual Meeting to consider a mechanism to address small cetaceans in the IWC. Following this, the IWC adopted a Resolution that developed a preliminary framework under which small cetacean issues can be addressed co-operatively in the IWC, notwithstanding the different views over legal competence (IWC, 1994, pp.15-16, 31-2). The Working Group has discussed the way stocks are identified for review by the Small Cetaceans Sub-committee of the Scientific Committee; coastal state participation (including non-member countries) in research and review; and improving availability and reliability of data and information. The IWC has also established a voluntary fund to assist participation by developing countries in these matters (IWC, 1995a, pp.20-1, 41-2).

The practical outworking of this situation is that the Scientific Committee can make *recommendations* to the Commission concerning the management of the great whales stocks which it investigates in terms of status and trends of abundance, while it provides *advice* on the smaller cetacean species.

CURRENT MANAGEMENT PRACTICES

The starting point for any discussion of the current activities of the IWC must be the present ban on commercial whaling world-wide. The difficulties in the management process and the uncertainty over the status of the stocks of great whales were important factors in the 1982 decision of the IWC to implement a pause in commercial whaling with effect from the 1986 coastal and 1985/86 pelagic whaling seasons. The IWC also agreed that by 1990 at the latest it would undertake a comprehensive assessment of the effects of the decision on whale stocks and consider modification and the establishment of other catch limits (IWC, 1983, pp. 20-1).

Comprehensive assessment
Following this decision, the Scientific Committee of the IWC embarked on the process which came to be known as the Comprehensive Assessment of whale stocks. It was defined by the scientists as an in-depth evaluation of the status and trends of all whale stocks in the light of management objectives and procedures (Donovan, 1989). Detailed assessments of gray whales in the North Pacific, bowhead whales off Alaska, minke whales in the Southern Hemisphere, North Atlantic and

western North Pacific, and North Atlantic fin whales have been carried out so far. Included in this process is the determination of current numbers of whales in the different stocks. The Scientific Committee has also accepted some other analyses, notably for humpback and pilot whales in the North Atlantic; the numbers which the IWC is prepared to give as the best estimates at this time are shown in the Annex.

An important aspect of this assessment programme has been the development of rigorous sighting survey and photo-identification techniques, coupled with robust methods of analysis of the resulting data, leading to statistically reliable estimates of abundance of the various whale stocks considered (e.g. Hiby & Hammond, 1989: Buckland & Duff, 1989). The IWC has now decided that only surveys and analyses of data which fulfil the guidelines drawn up by the Scientific Committee can be used to provide abundance estimates as input in implementing the Revised Management Procedure (IWC, 1996, pp.27, 45-6).

Revised management procedure
Another part of the comprehensive assessment programme was the development of a revised management procedure. Five different procedures were developed and tested by a series of computer simulation trials but it was not possible to complete this work by the original deadline of 1990 (Kirkwood, 1992). However, through the process of consultations and comparisons between the individual developers, these proposals came to the point where at its 1991 Annual Meeting the Scientific Committee recommended the IWC to adopt one procedure as suitable for implementation as the replacement for the "New Management Procedure" (NMP) which had been introduced in 1975 (IWC, 1992, pp.55-6).

The IWC formally adopted this revised procedure with some modifications at its 1994 Annual Meeting as the specification for the calculation of catch limits for baleen whales in a *Revised Management Procedure (RMP)* (IWC, 1995a, pp.26, 43-4). The objective is to provide an acceptable balance between conservation and exploitation of baleen whales and to provide a simple and convenient method for determining catch limits with minimal requirements for data. The procedure seeks to ensure that depleted stocks are rehabilitated, and that no whaling is permitted on stocks which are below 54% of their initial abundance. The aim is to obtain the highest possible continuing yield, with stable catch limits, to bring all stocks to the target level of 72% of their initial level. Both of these percentage levels have some relationship to the stock classifications based on the Maximum Sustainable Yield (MSY) concept which formed the basis of the now discredited NMP.

The only data input required for the revised procedure are a current population estimate and the known catch history. The notional unexploited stock size is estimated using a simplified production model that includes no biological parameters. Initially a fixed MSY rate is assumed, but as more data accumulate the procedure tends gradually to one based on the best estimate of the MSY rate obtained by fitting the production model. The procedure has been tested in simulation trials over 100 years and is robust to a range of factors, including underestimation of historic catches by up to 50%, variations over time in carrying capacity and recruitment, environmental degradations and a wide range of uncertainty, including stock units and differing population dynamics. It is also a very conservative regime.

Supervision and control
Although the scientific component of any all-embracing revised management scheme has been completed, the Commission has resolved that until all aspects of the Revised Management Scheme, including an effective inspection and observation scheme which fully addresses *inter alia* the issues of under-reporting and mis-reporting of catches, are incorporated into the Schedule the Revised Management Procedure should not be implemented (IWC 1995a, pp. 26, 43-4).

Data have recently been presented correcting the catch statistics submitted from Soviet whaling operations in the Southern Hemisphere from 1949/50 (Zemsky, Berzin, Mikhalyev & Tormosov, 1995). The earlier official figures submitted by the USSR were substantially falsified to conceal large scale violations of the international regulations. Such evidence has reinforced the need for the establishment of totally credible inspection and international observation schemes for any future whaling activities. 1972 saw the implementation of the IWC's International Observer Scheme, whereby observers appointed by the IWC and reporting directly to it were stationed at the whaling operations of the member states to confirm their compliance with the agreed whaling regulations (IWC, 1974, pp. 27-8). The quality of the Soviet official records improved after this time when compared with the original records collected on the whaling vessels.

The IWC is currently discussing an appropriate framework for new supervision and control procedures. A Working Group which met in Mexico in May 1994, Norway in January 1995, Ireland in May 1995, and the UK in June 1996, continues to try and resolve the various issues involved, which include consideration of questions of national sovereignty, independent verification, and reporting requirements.

Some governments would like international inspection and controls to extend from the initial surveys and analyses on which whale stock estimates are based, through the actual catching and processing operations at sea or on shore, to the distribution of the products in the marketing phase of the industry. Others believe that a less comprehensive approach is required, and in particular that the trade aspect is more properly the responsibility of the Convention on International Trade in Endangered Species of Wild Fauna and Flora (IWC, 1996, p.26).

Revision of the convention
A Convention signed in 1946 is clearly a reflection of attitudes and understanding which may well have developed and changed over the intervening decades. Even before the recent adoption in the context of the United Nations Convention on the Law of the Sea of the current concepts of coastal state sovereignty and the precautionary principle of management, there had been profound changes in the interpretation and application of the 1946 whaling Convention and the consequent management policies by which it is implemented.

There have been three major attempts to revise the 1946 Convention in the past 20 years. In 1974 the USA submitted a draft protocol to amend the Convention and the Commission established a Working Group of interested member nations to discuss the problems posed to the present Convention "Taking into account the changes which have occurred in whaling and stocks of cetaceans since 1946 and bearing in mind the necessity to strengthen the mechanism for the international conservation of whales and their rational management both at present and in the future; recognizing that the discussions in the Law of the Seas Conference may affect the activities of the IWC" (IWC 1976, p. 32).

A revised text and amendments proposed by different countries by postal communication were collated into a single document for discussion. This was the basic negotiating text considered at a preparatory meeting convened by the Government of Denmark in Copenhagen in July 1978 when alternative texts were developed (IWC 1980, p. 32).

Portugal (not a member of the IWC) then extended an invitation for a Drafting Group to meet in Estoril in November 1979, but this group was able to agree on a text for only the Preamble, the first and part of the second Articles (IWC 1981, p. 28).

A third meeting was hosted by the Government of Iceland in Reykjavik in May 1981 to improve and update the Convention. This had been rather more successful because the discussions had not been concerned with wording but the principles involved. However, there was still no overall consensus on the key issues of the object and purpose of the Convention, jurisdiction (species and areas), membership and decision taking. The Commission agreed to leave the matter as it was for the time being (IWC, 1982, p.34).

CURRENT INTERPRETATIONS

The tensions between the objectives of the conservation of the whale resources and the orderly development of the whaling industry continue today. Some governments, such as Iceland, Japan and Norway, argue for a resumption of commercial whaling as an example of sustainable development of the resource now that the scientific aspects of a revised management procedure have been agreed. Others, including Australia, France, Germany, Netherlands, UK and USA, taking a more protectionist position and are reluctant to do anything which could lead to a repetition of the past over-catching of whales.

The movement towards restriction, and eventual prohibition, of all whaling can be traced very clearly through the meetings of the IWC since 1972. In that year the UN Conference on the Human Environment met in Stockholm, Sweden and adopted a three point Resolution recommending the strengthening of the IWC, increased international research efforts, and calling for an agreement involving all governments concerned in a ten year moratorium on commercial whaling. The proposal for a global moratorium was rejected by the Commission at its meeting, but actions to implement the other two elements were approved (IWC, 1974, pp.23-6).

The individuals, organisations and governments active in working for a cessation of whaling over the following years gradually achieved the introduction of progressively lower catch limits, the prohibition of pelagic whaling except for minke whales and the establishment of the Indian Ocean Sanctuary in 1979 (IWC, 1980, pp.26-7), and finally the setting of zero catch limits for all commercial whaling adopted in 1982 with effect from 1986 (IWC, 1983, pp.20-1).

Since achieving this position, there has been a reluctance to consider any resumption of commercial whaling despite appropriate scientific justification in terms of stock abundance estimates and the development of the Revised Management Procedure, on the grounds that other factors must also be settled first. These include adequate inspection and observation schemes, discussed above, and humane killing. Indeed, the Commission has since put in place the Southern Ocean Sanctuary as a further limit on commercial whaling (IWC, 1995a, pp. 27-9).

Whale sanctuaries
As well as management measures governing catch and size limits, species and seasons, the IWC may also designate open and closed areas for whaling. Even before the 1946 Convention a Sanctuary in the Antarctic was established by the majority of whaling nations in 1938 south of 40°S between longitudes 70°W and 160°W. The original reason for this was that in this sector commercial whaling had not hitherto been prosecuted and it was thought highly desirable that the immunity which whales in this area had enjoyed should be maintained. This sanctuary was continued by the IWC from its inception until 1955, when the area was opened initially for three years and then continuously as a means of reducing the pressure of catches on the rest of the Antarctic whaling grounds (Tønnessen & Johnsen, 1982, pp. 563-5).

The Indian Ocean Sanctuary was established by the IWC in 1979, extending south to 55°S latitude, as an area where commercial whaling is prohibited. This provision was an initiative of the newly joined Government of the Seychelles to provide freedom from disturbance for ecosystems and for species or groups of animals, especially for activities such as breeding. The Seychelles stated that its proposal satisfied the ecological coherence of the area and the expressed opinions of several neighbouring countries. (IWC, 1980, p. 27). The Indian Ocean Sanctuary was initially established for 10 years and its duration has since been extended twice.

It will be reviewed again by the IWC at its Annual Meeting in the year 2002 (IWC, 1993, p. 27).

At its 1992 Meeting the IWC received a proposal from France for the establishment of a whale sanctuary in the Southern Hemisphere south of 40°S. The purpose of the proposal was stated to be to contribute to the rehabilitation of the Antarctic marine ecosystem and the protection of all Southern Hemisphere species and populations of baleen whales and the sperm whale on their feeding grounds. This would also link up with the Indian Ocean Sanctuary to provide a large area within which whales would be free from commercial catching (Government of France, 1993). Following debate, France decided to defer consideration of the proposal to the 1993 Annual Meeting.

An IWC Working Group examined the matter over the next two years and at the 46th (1994) Annual Meeting the IWC adopted the Southern Ocean Sanctuary as another area in which commercial whaling is prohibited. The northern boundary of this Sanctuary follows the 40°S parallel of latitude except in the Indian Ocean sector where it joins the southern boundary of that sanctuary at 55°S, and around South America and into the South Pacific where the boundary is at 60°S. This prohibition will be reviewed ten years after its initial adoption and at succeeding ten year intervals, and could be revised at such times by the IWC (IWC, 1995a, p. 28).

Humane killing
A number of governments are expressing increasing concern over the question of humane killing in the whaling context. The IWC has urged that its members continue to promote the development of humane killing methods. During the past ten years both Japan and Norway have made major improvements in their whale killing technology. This includes advances in the triggering mechanism for the explosive grenades used in the minke whale hunts, and the introduction of an improved explosive material, penthrite, which has greater power than the traditional black powder. These improvements are such that they are now also being introduced into the aboriginal subsistence whaling hunts in Alaska and Greenland (IWC, 1995a, pp. 15-16). Any assessment of the efficiency, usually measured by the time to death in the whaling operations, relies on a rather subjective indication of insensibility and death. There is still a need to relate the cause of death to the observed time to death, as well as the collection of information on the physiological status of hunted animals.

A Workshop on Whale Killing Methods was held in the UK in June 1992 and was re-convened in Ireland in May 1995 to review and evaluate progress on the action plan it adopted, and to review the use of the electric lance as a secondary method for killing whales. A proposal to ban the use of the electric lance, which is employed as a secondary killing method in the Japanese Antarctic minke whale fishery, failed to receive a sufficient majority when it was discussed at the 1996 Annual Meeting of the Commission because of the conflicting evidence, and a Working Group to examine the scientific issues will meet in 1997 (IWC, 1997).

THE FUTURE

Whalewatching
A major new development is the IWC's involvement in whalewatching as a sustainable use of cetacean resources. In 1993 the IWC invited Contracting Governments to undertake a preliminary assessment of the extent, and economic and scientific value of whalewatching activities. These reports on the value and potential of whalewatching were consolidated by the Secretariat and considered by a Working Group at the 1994 meeting (IWC, 1995a, p. 32). As a result the IWC has reaffirmed its interest in the subject, encouraged some scientific work and adopted a series of objectives and principles for managing whalewatching proposed by the Scientific Committee. In 1996 it adopted a Resolution which underlined the IWC's future role in monitoring and advising on the subject (IWC, 1997).

The environment and whale stocks
With increasing awareness that whales should not be considered apart from the marine environment which they inhabit, and that detrimental changes may threaten whale stocks, the IWC decided that the Scientific Committee should give priority to research on the effects of environmental changes on cetaceans. The Scientific Committee examined this issue in the context of the Revised Management Procedure and agreed the RMP adequately addresses such concerns. However, it went on to state that the species most vulnerable to such threats might well be those reduced to levels at which the RMP, even if applied, would result in zero catches (IWC, 1995b, pp. 55-9). The Scientific Committee has held two workshops, one on the effects of chemical pollutants in March 1995 in Norway, and another on the effects of climate change and ozone depletion in March 1996 in the USA. The IWC has adopted Resolutions reaf-

firming its view of the importance of these matters and encouraging the Scientific Committee to increase collaboration and cooperation with governmental, regional and other international organisations working on related issues (IWC, 1997).

OTHER ASPECTS

North Atlantic Marine Mammal Commission
Iceland and Norway have stated in recent meetings of the IWC that since that body seems reluctant to allow the resumption of commercial whaling, it might be necessary for them to turn to alternative fora where the issues of catch limits can be discussed. Iceland in fact withdrew from the Convention in 1992. In this context the newly established North Atlantic Marine Mammal Commission, comprising government representatives from the Faroe Islands, Greenland, Iceland and Norway is relevant. The United Nations Convention on the Law of the Sea does not specify which international body, or even if there is to be only one body, is the appropriate organisation through which coastal states are to co-operate with respect to the management and conservation of the whale resources (Article 65). However, the Agenda 21 Oceans Chapter adopted at the United Nations Conference on Environment and Development held in Rio de Janeiro in 1992 does recognise the responsibility of the IWC for this role, whilst not precluding regional agreements (paragraph 17.90).

Certainly the precedent for regional bodies to oversee the whaling activities of coastal states in particular parts of the world is not new. The governments of Chile, Ecuador and Peru formed the Permanent Commission of the Conference on the Use and Conservation of the Marine Resources of the South Pacific in 1952. They established catch regulations broadly similar to those in force for the IWC at that time. Subsequently all those governments have joined the IWC - Chile and Peru in 1979, and Ecuador in 1991, although the latter withdrew again in 1994.

Scientific research catches
The 1946 Convention allows any member government to grant a special permit for the taking of whales for scientific research purposes. In response to proposals to issue permits since the moratorium decision on commercial whaling the IWC has established detailed guidelines for its Scientific Committee to review and comment on such proposed permits

(IWC, 1987, p.25; 1988, p.27; 1996, pp. 46-7). The IWC may also make recommendations to the governments concerned although it has no power to forbid or rescind the issue of a permit.

Iceland, while it was still a member of the IWC, carried out a four-year programme of research on the biology, genetics and ecological role of the whales in its waters from 1986 to 1989, taking 292 fin whales and 70 sei whales. The Commission passed Resolutions asking Iceland to reconsider and refrain from this activity (IWC, 1988, p.28; 1989, pp.30-1; 1990, p. 35).

Norway carried out a seven-year research programme started in 1988 to study and monitor minke whales in the Northeast Atlantic, including investigations on feeding ecology, age determination and aspects related to energetics. This was part of a broader ecological programme designed to provide information for future multi-species management in the Barents Sea. The IWC has regularly adopted Resolutions each year inviting Norway to reconsider the lethal takes (e.g. IWC, 1989, p.30; 1995a, p.48), which amounted to 289 whales.

Following two years of feasibility studies starting in the 1987/88 season to resolve the problems of collecting representative samples, Japan has embarked on a long-term research programme in the Antarctic which includes an annual catch of 300 or 400 ± 10% minke whales to estimate the biological parameters which could be used for management, in particular natural mortality, and the elucidation of the role of whales in the Antarctic ecosystem. The IWC again regularly adopts a Resolution each year inviting Japan to reconsider this research programme (e.g. IWC, 1988, p.29; 1997), to which the Japanese authorities always give serious consideration and a reasoned response.

Japan also initiated in 1994 a programme to clarify the stock structure and mixing rates of minke whales around its coasts, based on problems encountered by the Scientific Committee in trying to assess the Northwestern Pacific stocks. The take is of 100 animals each year, and the IWC regularly asks Japan to reconsider this programme (IWC, 1995a, p. 47; 1997).

Many countries over the years have issued permits to allow limited catches of whales which would not otherwise be sampled in the normal commercial operations, such as lactating females, under-sized animals or protected species, in order to advance scientific knowledge. However, one aspect of the current research catches which has created some concern is the requirement in Article VIII paragraph 2 of the 1946 Convention that the whales taken "shall so far as practicable be processed". In the situation that commercial whaling is currently prohibited, the suspicion has been that the present catches are a way of circumventing the ban or of maintaining the whaling operations until there is a full resumption. Whatever is the truth, it has to be said that useful results from these research activities are being presented to the Scientific Committee, even if the objectives of some of the research fall outside the guidelines adopted by the IWC.

CONCLUSION

Commercial whaling has a long history of over-exploitation of each species and stock of whales as they were discovered or the technology advanced to permit capture. No regulations have been successful in preventing this in the face of the economic pressures from the industry.

The current scientific work on the comprehensive assessment of whale stocks and a revised management procedure have now reached the point where catch limits for commercial whaling on certain stocks can be calculated using the agreed procedure which has been rigorously tested by computer simulation, and requires a minimum of data. But the IWC has not yet considered setting catch limits other than zero, because there are still concerns to be satisfied over the safeguards, including adequate inspection, international observation, data reporting and monitoring, which are considered necessary by some nations to be built into a full revised management scheme. The problems of humane killing also remain to be resolved.

The uncertainties involved in whale stock assessment, management and regulation have only served to reinforce the views of those people and nations who are opposed to a resumption of commercial whaling of the need for a large sanctuary to provide a pool of whales which are sure to be safe from hunting.

Against this position it is argued by governments such as Norway, which has lodged objections to the ban on commercial whaling and licences its nationals to take limited catches, that some whale stocks undoubtedly could sustain carefully regulated and controlled catches. In addition, and as discussed in another paper to this conference, certain coastal communities would like to resume their traditional whaling activities which are at present banned by the moratorium on commercial whaling. These communities have society structures and traditions similar to others which are allowed to continue hunting whales under the IWC's aboriginal subsistence whaling provisions.

Public opinion in some parts of the developed world seems increasingly to regard cetaceans as 'special' animals which should not be hunted for the products which they yield but seen more as a focus for whalewatching, educational programmes and aesthetic values. This gives rise to a clash of cultures with hunting communities concerning the responsible use of this resource.

It may be that the final solution will be to establish total protection from hunting for whales throughout the greater part of the world's oceans, but to permit limited and carefully monitored catches by traditional and cultural operations within the 200 nautical miles EEZs of those countries with communities wishing to maintain and continue these activities. Such a policy may well reflect the economic realities as well as the aspirations of both the consumers and the conservation interests in today's world (Gambell, 1990, pp.41-2). Given the flexibility of interpretation of the 1946 Convention which has occurred over the past half century, it is not certain that a new Convention would necessarily be needed to enshrine such a policy in international law. It really depends on the collective will of the member nations of the IWC as they face this particular challenge of the new millennium.

REFERENCES

Agenda 21 Oceans Chapter, Rio de Janeiro, 1992. Pp. 1095-1120 in *The Marine Mammal Compendium of Selected Treaties, International Agreements, and Other Relevant Documents on Marine Resources, Wildlife, and the Environment.* Marine Mammal Commission, Washington, D.C. USA.

BUCKLAND, S. T. & Duff, E. I. 1989. Analysis of Southern Hemisphere Minke Whale Mark-Recovery Data. *Rep. Int. Whal. Commn* (Special Issue 11):121-143.

DONOVAN, G. P. 1989. Preface. The Comprehensive Assessment of Whale Stocks: the early years. *Rep. Int. Whal. Commn* (Special Issue 11).

GAMBELL, R. 1990. The International Whaling Commission -quo vadis? *Mammal Rev.* 20(1):31-43.

GAMBELL, R. 1995. Management of whaling in coastal communities. Pp. 699-708 in *Whales, Seals, Fish and Man* ed. Blix, A. S., Walløe, L. & Ulltang, Ø. Elsevier, Amsterdam.

Government of France. 1993. A Southern Ocean Whale Sanctuary. *Rep. Int. Whal. Commn* 43:41-48.

HIBY, A. R. & Hammond, P. S. 1989. Survey Techniques for Estimating Abundance of Cetaceans. *Rep. Int. Whal. Commn* (Special Issue 11):47-80.

IWC 1974. Chairman's Report of the Twenty-Fourth Meeting. *Rep. Int. Whal. Commn* 24:20-36.

IWC 1976. Chairman's Report of the Twenty-Sixth Meeting. *Rep. Int. Whal. Commn* 26:24-40.

IWC 1980. Chairman's Report of the Thirty-First Annual Meeting. *Rep. Int. Whal. Commn* 30:25-41.

IWC 1981. Chairman's Report of the Thirty-Second Annual Meeting. *Rep. Int. Whal. Commn* 31:17-40.

IWC 1982. Chairman's Report of the Thirty-Third Annual Meeting. *Rep. Int. Whal. Commn* 32:17-42.

IWC 1983. Chairman's Report of the Thirty-Fourth Annual Meeting. *Rep. Int. Whal. Commn* 33:20-42.

IWC 1988. Chairman's Report of the Thirty-Ninth Annual Meeting. *Rep. Int. Whal. Commn* 38:10-31.

IWC 1989. Chairman's Report of the Fortieth Annual Meeting. *Rep. Int. Whal. Commn* 39:10-32.

IWC 1990. Chairman's Report of the Forty-First Annual Meeting. *Rep. Int. Whal. Commn* 40:11-37.

IWC 1991. Chairman's Report of the Forty-Second Meeting. *Rep. Int. Whal. Commn* 41:11-50.

IWC 1992. Report of the Scientific Committee. *Rep. Int. Whal. Commn* 42:51-136.

IWC 1993. Chairman's Report of the Forty-Fourth Annual Meeting. *Rep. Int. Whal. Commn* 43:11-53.

IWC 1995a. Chairman's Report of the Forty-Sixth Annual Meeting. *Rep. Int.*

Whal. Commn 45:15-52.
IWC 1995b. Report of the Scientific Committee. *Rep. Int. Whal. Commn* 45:52-103.
IWC 1996. Chairman's Report of the Forty-Seventh Annual Meeting. *Rep. Int. Whal. Commn* 46:15-48.
KIRKWOOD, G. P. 1992. Background to the Development of Revised Management Procedures. *Rep. Int. Whal. Commn* 42:236-243.
TØNNESSEN J. N. & Johnsen, A.O. 1982. *The History of Modern Whaling.* xx+798 pp. C. Hurst & Co. London.
United Nations Convention on the Law of the Sea. 1982. United Nations, New York.
ZEMSKY, V. A., Berzin, A. A., Mikhalyev, Yu. A. & Tormosov, D. D. 1995. *Materials on Whaling by Soviet Whaling Fleets* (1947 - 1972). 320 pp. Moscow.

ANNEX - WHALE POPULATION ESTIMATES
(rounded to the third significant figure of the upper confidence limit)

Because of the considerable scientific uncertainty over the numbers of whales of different species and in different geographical stocks the International Whaling Commission decided in 1989 that it would be better not to give whale population figures except for those species/stocks which have been assessed in detail, and for which there is statistical certainty with respect to the numbers.

Population	Year	Size	95% Confidence interval
MINKE WHALES			
Southern Hemisphere	**1982/83 - 1988/89**	761,000	510,000 - 1,140,000
North Atlantic (excluding Canadian East Coast area)	**1987-95**	Approx. 149,000 (subject to slight revision)	Approx. 120,000 - 182,000
North Pacific (North West Pacific and Okhotsk Sea)	**1989-90**	25,000	12,800 - 48,600
FIN WHALES			
North Atlantic	**1969-89**	47,300	27,700 - 82,000
SEI WHALES			
Central North Atlantic	**1989**	10,300	6,100 - 17,700
GRAY WHALES	**1987/88**	21,000	19,800 - 22,500

Assuming a constant rate of increase, the population was increasing at a rate of 3.2% (95% confidence interval 2.4% - 4.3%) over the period 1967/68 - 1987/88 with an average annual catch of 174 whales.

BOWHEAD WHALES
Bering-Chukchi-Beaufort Seas stock **1988** 7,500 6,400 - 9,200

The net rate of increase of this population from 1978 to 1988 has been estimated as 2.3% per year (95% confidence interval 0.9% - 3.4%).

HUMPBACK WHALES
Western North Atlantic **1979-86** 5,500 8,120 - 2,890

A rate of population increase of 10.3% (95% confidence interval 2% - 23%) was obtained from the Gulf of Maine.

BLUE WHALES
Southern Hemisphere **1985/86-1990/91** 460 210 - 1,000

PILOT WHALES
Central & Eastern North Atlantic **1989** 780,000 440,000 - 1,370,000

ABSTRACT

The 1946 International Convention for the Regulation of Whaling was drawn up to control the catching operations of a particular fishery. It embodies such current ideals as wise use and sustainable development, and established the International Whaling Commission which can amend the whaling regulations. As the term 'whale' is not defined. some governments believe the Convention applies to all cetaceans, while others restrict it to just the great whales. However, it is agreed that the Scientific Committee can provide advice on the small cetaceans.

Since 1986 the IWC has banned commercial whaling while the Scientific Committee carries out a comprehensive assessment of the status and trends of the whale stocks. This has seen the development of rigorous survey techniques and analyses, and a Revised Management Procedure for baleen whales. Before this is implemented the Commission has to agree on effective monitoring, inspection and observation schemes.

Attempts to revise the Convention have failed due to differences of opinion over the purpose and scope of a new instrument. In the meantime the provisions of the present Convention have been used to progressively lower catch limits for commercial whaling to zero, establish the Indian Ocean and Southern Ocean Sanctuaries, and promote more humane killing methods. New interests now include whalewatching and environmental impacts on whale stocks.

The formation of an alternative body by nations wishing to resume whaling and the continuation of small catches for scientific research purposes puts pressure on the IWC to accommodate the views of people who have a tradition of sustainable harvesting in coastal waters as well as those who regard whales as animals to be protected. It may be that both positions can be achieved under the present Convention.

The North Atlantic Marine Mammal Commission - in principle and practice

Kate Sanderson *Secretary to NAMMCO, Søndre Tollbodgate 9, University of Tromsø, 9037 Tromsø, Norway*
e-mail: Kate.Sanderson@nammco.no

NAMMCO IN CONTEXT -
BACKGROUND, ESTABLISHMENT AND STRUCTURE

The North Atlantic Marine Mammal Commission was formally established with the signing of the Agreement on Cooperation on Research, Conservation and Management of Marine Mammals in the North Atlantic. This was done on April 9 1992 in Nuuk, Greenland by the fisheries ministers of the Faroe Islands, Greenland, Iceland and Norway.

The year 1992 is also remembered for certain other events on the international front related to discussions on whaling in the North Atlantic; namely Iceland's withdrawal from the International Whaling Commission, and Norway's announcement the same year that commercial minke whaling would be resumed in 1993. On the broader international arena, 1992 was also the year of the UN Earth Summit in Rio, the follow-up to the World Commission on Environment and Development, and a reaffirmation of the principles of conservation and sustainable development.

In this context, then, the NAMMCO agreement can be seen as very much a product of its times, reflecting a recognition of the need to explore new approaches to international cooperation on conservation and management and a desire for the meaningful application of the general principles at stake in terms of resource conservation and sustainable use. More specifically, one of the prime motivating factors behind the creation of NAMMCO was the dissatisfaction in the North Atlantic with the inability of the IWC to agree on a basis for conservation and management of large whales according to these principles and its own Convention.

The political process leading to the establishment of NAMMCO goes back to 1988, when the first of a series of annual international conferences on the rational utilization of marine mammals was hosted by the Icelandic minister of fisheries at the time, Halldór Ásgrímsson. Parties to the Conference were the present four NAMMCO member countries as well as Canada, Russia and Japan.

From the Conference forum there developed a memorandum of understanding which was signed in 1990 between the Faroes, Greenland, Iceland and Norway establishing the North Atlantic Committee on Cooperation on Marine Mammal Research (NAC), which paved the way for the formal development of the NAMMCO agreement and the establishment of this new body in 1992.

Despite the obvious causal link between developments in the IWC and the political genesis of NAMMCO, the future significance of NAMMCO as an appropriate international regime is found in those aspects of the agreement which break new ground and provide new emphases for managing marine resources, within a general framework which is fully consistent with requirements for international cooperation on marine mammals under the UN Convention on the Law of the Sea.

Firstly, NAMMCO covers all marine mammals in the North Atlantic, including smaller species of whales as well as seals and walruses for which there has not previously existed an international mechanism for cooperation on their conservation and management.

Perhaps most important is the conviction of the Parties to the Agreement, highlighted in the preamble, that regional bodies in the North Atlantic can ensure effective conservation and sustainable marine resource utilization and development with due regard to the needs of coastal communities and indigenous people. In this respect, NAMMCO is itself the end result of a prevailing trend towards regional rather than global approaches to conservation, motivated by a desire to reduce the distance between resource managers and resource users, and ensure an effective consultation process so that local communities are involved in conservation and management decisions which may ultimately affect their lives and livelihoods.

At the cornerstone of conservation is the need for reliable scientific knowledge of resources on which to base management decisions. Built into the NAMMCO agreement is the desire for a better understanding of the dynamics of the marine ecosystem - the ultimate producer upon which coastal communities rely. Research collaboration through NAMMCO, with a starting point in specific marine mammal species and stocks of interest to member countries, focuses at the same time on the complex ecological role of marine mammals as both predators and prey, as potential competitors for resources, and, not least, as potential victims of human impacts such as pollution in the marine ecosystem.

The structure of NAMMCO is based closely on the model of another regional body familiar to the founders, namely NASCO. The Commission is made up of a Council, which provides the main forum for the exchange of information among Parties, establishes Management Committees and guidelines for their work, and coordinates requests for scientific advice. Management Committees (at present in the form of a general Management Committee) propose to their members measures for conservation and management, and make recommendations to the Council concerning scientific research. The Scientific Committee, to which each member country appoints three scientific experts, provides scientific advice in response to requests from the Council. A number of other Working Groups have also been set up under the Commission, which are described briefly below. The Secretariat was established in Tromsø, Norway in September 1993 and currently has a staff of three.

Decisions of the Council and Management Committees must be made by consensus. The Agreement is open for signature by other Parties with the consent of existing Signatories.

NAMMCO IN PRACTICE - THE FIRST FIVE YEARS

As a basis from which to begin its work, the Scientific Committee compiled a selected list of those marine mammal species with direct or indirect management interest to members of the Commission. Requests for advice on a number of the following species and stocks have been forwarded by the Council to the Scientific Committee, which has dealt with these at its annual meetings since 1993:

Killer whales - ongoing
Northern bottlenose whales - stock assessment & population modelling
Atlantic walrus - assessment across range
Ringed seal - assessment across range
Grey seal- assessment across range & focus on sealworm
Pilot whale- effects of Faroese drive hunt (through ICES Study Group)
Harp & hooded seals- assessment across range   (Joint ICES/NAFO Working Group on Harp and Hooded Seals)

Built into the Council's request for advice on these species, apart from questions related to abundance and distribution, is also a focus on their role in the ecosystem, including interactions with other marine resources, the effects of environmental changes and changes in the food supply.

With the exception of the killer whale, the Scientific Committee has completed assessments of the species listed above and has presented its findings to the Council. In the case of the pilot whale and harp and hooded seals, NAMMCO's requests for advice on these species have been dealt with by existing expertise in other fora. The Study Group on Pilot Whales established by ICES has now completed its work, and its findings will be reviewed by the Scientific Committee at its 1997 meeting in relation to the Council's request for an evaluation of the sustainability of the Faroese hunt. A number of issues are outstanding in the work of the ICES/NAFO Joint Working Group on Harp and Hooded Seals, but population assessments for the Northwest Atlantic were available in 1996 and reviewed by the Scientific Committee at its 1996 meeting. As for the grey seal, further assessment of the dynamics of the sealworm has been requested, and the Scientific Committee will deal with this at its 1997 meeting, to which a number of experts in this field have been invited.

Important for the Scientific Committee is the fact that sufficient funding is earmarked to ensure the participation of relevant experts in its work. The recent comprehensive assessments of the Atlantic walrus, ringed seals and grey seals, for example, would not have been possible without the valuable active contribution of scientists from a number of other, non-member countries, in particular from Canada, but also from the UK and Russia.

Based on the findings of the Scientific Committee, the general Management Committee has agreed on a number of proposals for conservation and management at its last two annual meetings, summarised as follows:

Northern bottlenose whale: coastal use of bottlenose whales in the Faroe Islands is sustainable - catch of up to 300 whales a year would not lead to stock decline;
Atlantic walrus: recommend measures to arrest decline in stock along west coast of Greenland;
Ringed seal: present catch levels in West Greenland / Canada are sustainable;
Harp seals: 1990-95 catch levels in Northwest Atlantic are well below replacement yields;
Hooded seals: 1990-95 catch levels in Northwest Atlantic are below replacement yields

Other ongoing work in the Scientific Committee includes the continued monitoring of stock levels and trends in stock levels of all marine mammals in the North Atlantic. The Scientific Committee Working Group on Abundance Estimates, which met in Reykjavik in February 1997, has been reviewing the results of the 1995 North Atlantic Sightings Survey (NASS 95). NASS-95 was coordinated by the NAMMCO Scientific Committee and carried out in the summer of 1995 through national whale sightings surveys in Iceland, Norway and the Faroe Islands. The report of the Abundance Estimates Working Group, which will be presented to the Scientific Committee at its next meeting, will provide an overview of the latest information on the abundance of whales in the North Atlantic, including updated estimates for the Central Atlantic minke whale stock, as well as fin, sei and pilot whales, based on the data from NASS-95 in relation to data from NASS surveys carried out in previous years.

Also on the agenda for the next Scientific Committee meeting is a closer look at the role of marine mammals in the ecosystem with a specific focus on the food consumption of three major predators in the North Atlantic - the minke whale, the harp seal and the hooded seal. As mentioned previously, there will also be a closer review of sealworm infestation in North Atlantic coastal areas, as a follow-on from the 1996 assessment of the status of the grey seal in the North Atlantic.

One other important new development through NAMMCO has been the adoption at the 1996 Council meeting of the Joint NAMMCO Control Scheme for the Hunting of Marine Mammals, which was developed by a Working Group set up under the Management Committee. A central precondition for international cooperation on management of shared resources is certainty that management decisions are upheld and respected in member countries. The NAMMCO Control Scheme provides a mechanism through which NAMMCO members can maintain international transparency in their utilisation of marine mammals.

The Control Scheme has two main elements. The first is an outline of agreed common elements for national inspection schemes specific to whaling operations with a harpoon cannon. The second element is the International Observation Scheme, with application to all types of whaling and sealing in the NAMMCO area, providing for an exchange of observers appointed by NAMMCO. The role of observers, who have no jurisdiction over the activities they observe, is to report on their observations to NAMMCO through the Secretariat.

The Joint Control Scheme is expected to be implemented to some extent, at least on a trial basis, already in 1997.

Other issues of interest to the Council include the levels and effects of contaminants in marine mammals. NAMMCO arranged an International Conference on Marine Mammals and the Marine Environment which was held in Shetland in 1995, and which focused on the sources, levels and effects of pollutants in marine mammals, as well as the health implications of contaminants for coastal communities where whales and seals are an important part of the diet. Papers presented at this Conference have now been published as a Special Issue of the journal *The Science of the Total Environment* (Elsevier, Vol. 186, 1-2, 1996).

To conclude this brief overview of NAMMCO's work, mention should also be made of the Working Group on Hunting Methods, which is a forum for the exchange of technical information and advice on methods used in whaling and sealing in member countries; and the NAMMCO Fund, which was set up as a source of funding for information projects related to the conservation and sustainable use of marine mammals.

It was decided last year that funds earmarked for the NAMMCO Fund in 1996 should be used to focus on seals and sealing. As a result, it has now been decided that NAMMCO will arrange an international conference on seals and sealing, and this will be co-sponsored by the Inuit Circumpolar Conference, the Nordic Council of Ministers, the Nordic Atlantic Cooperation and High North Alliance. The provincial Government of Newfoundland and Labrador in Canada has offered to host the event, which will be held in St John's from 25-27 November 1997.

NAMMCO IN THE FUTURE - PROGRESS AND PROSPECTS

To draw some conclusions on the work of NAMMCO so far, it would have to be said that the organisation has come a very long way in short space of time, in particular through the work of the Scientific Committee.

In general terms it could also be said that one of the strengths of NAMMCO is the simplicity and therefore flexibility of the agreement and the fact that all decisions must be made by consensus. This requires a real interest and political will on the part of member countries to apply the principles of the Agreement in practice in a way which is acceptable to all Parties, and which takes account of all relevant interests represented in the forum.

How much further NAMMCO will be used actively for management advice and recommendations on other species and issues in the future is entirely a matter for the member countries to agree on. This will depend on the prevailing international political climate and the level of conviction which inspired the creation of the organisation in the first place. The principles are agreed, the organisation has so far operated effectively and has begun to deal with issues which have a practical bearing on management policies and the interests of resource users in member countries.

The long-term agenda has still to be drafted.

Additional membership would obviously also shape NAMMCO's future agenda. The Council of NAMMCO has invited the governments of Canada and the Russian Federation to become full Parties to the NAMMCO Agreement. This option is currently being considered in Ottawa and Moscow, and some indication of views is expected before the next meeting of the Council in Tórshavn, Faroe Islands in May 1997.

NAMMCO, IWC and the Nordic Countries

Steinar Andresen *The Fridtjof Nansen Institute, P.O. Box 326, N - 1324 Lysaker, Norway e-mail: steinar.andresen@fni.no*

SCOPE AND PURPOSE

This is a short *policy* paper without much academic pretensions. It contains a lot of speculations and thoughts about the future of the whaling issue, linked especially to the role of NAMMCO. Although it is impossible to discuss NAMMCO without touching upon the role of the IWC, it will not be dealt with in detail. Moreover, I will not discuss the formalities, objectives, history etc. of both the IWC and the NAMMCO, as this is considered to fall outside the scope of this paper. Although this is a policy paper, it builds to some extent upon my previous research mainly within the project on "The Effectiveness and Implementation of International Environmental Commitments."[1] I have written a chapter in a forthcoming book from the project on "The Making and Implementation of Whaling Policies: Does Participation Make a Difference?" where I analyze i.a. the whaling policies of Iceland and Norway. A few broad observations from this study are utilized as a basis for the present discussion on NAMMCO.

Concerning the Nordic countries, Norway and Icelands will be emphasized as the two key players in this connection.[2] I will start out by discussing briefly the significance of the Nordic countries in international resource and environmental cooperation in general, before turning to

---

[1] This research project has been organized and financed by the International Institute for Applied Systems Analysis (IIASA) in Laxenburg, Austria. David Victor and Gene Skolnikoff, both MIT, were directors of the project, which lasted from 1994 to 1996.
[2] This is not to say that especially Greenland and the Faroe Islands are interesting players in this game;

their positions on the whaling issue. In the last section I will turn to NAMMCO; what has it achieved and what are the future perspectives?

THE NORDIC SETTING: RHETORIC AND REALITY

In meetings within the Nordic Council and at other similar Nordic events, emphasis is often placed on interests these countries share in international affairs, not least on environmental issues. Although this rhetoric is still quite common, in real terms it is more a bygone phenomenon. The Nordic countries could largely act as a block during the 1970s and most of the 1980s, when international environmental agreements were rather "toothless"; they were of a declaratory nature and had little or no practical effect for the member countries. There were few, if any, expenses connected with being "green". Considering the Nordic reputation for being rather advanced in this area as well as their inclination to support "good causes" this was an issue well suited to presentation as a common Nordic concern.

However, over the course of the last ten years or so, this has changed dramatically as many international environmental agreements have become much more concrete and demanding; real interests and real behaviour are often affected if the agreements are to be complied with. This has strongly affected the Nordic unity since they have different material interests in many respects, which often makes Nordic alliances more the exception than the rule. The picture is not idyllic regarding fisheries either; as most notably exemplified by the conflict between Norway and Iceland. This conflict is so profound that it seems to have some spill-over-effect to the whaling issue; it becomes difficult to cooperate in this area as long as the fisheries conflicts are unresolved. Reducing Nordic unity even more is the fact that Denmark, Sweden and Finland have become EU members while Norway and Iceland have not. While the EU is playing an increasing role as a decisionmaking arena, the Nordic arena is losing much of its significance. Nevertheless, what has been accomplished within the Nordic setting should not be forgotten, i.a. in relation to the labour market and the open borders. Moreover, the role of norms, culture and history should be remembered. In such a perspective, the Nordic arena may still have a function. It provides an opportunity for *discussion* and exchange of ideas, though other actors and arenas usually need to be involved when *decisions* on management issues are being taken.[3]

---

not least their interactions with Denmark. As I do not know very much about this issue, however, I will just touch upon it briefly.
[3] The setting for this seminar is a perfectly good example showing that the Nordic level may be a useful one - provided its limitations are properly understood.

THE NORDIC COUNTRIES AND THE WHALING ISSUE: A BROAD PICTURE

Whaling is considered both an environmental and a resource management issue. Everybody familiar with this issue over the last decade or so knows that many people conceive the whaling issue as being primarily environmental. This is captured in the following quote by one of the most prominent persons known for turning the whaling issue into an environmental one: "saving the whale is for millions of people a crucial test of their political ability to halt *environmental* destruction" (Holt 1985:12) (emphasis added). Thus, the NAMMCO members are at odds with the *majority* of IWC members in regarding whaling as a question of sustainable harvest of a living marine resource. The administrative basis of the whaling issue in the Nordic countries illustrates the split in how the issue is perceived. As pointed out by Ivarsson (1994: 180), in Iceland, Norway, Greenland and the Faroe Islands whaling issues are under the Fisheries Administration. In the other Nordic countries they are a part of the Environmental Administration.

There may not be many issues among the Nordic countries where the spread in positions and preferences is more widespread than on the whaling issue. In fact, the main positions of the whole body of members of the IWC are represented among the five Nordic countries. If a continuum ranging from protection, conservation, aboriginal whaling to commercial whaling is conceived, the following broad picture would emerge:

FINLAND	SWEDEN	DENMARK	(ICELAND) NORWAY
Protection	Conservation	Aboriginal whaling	Commercial whaling

Both Finland and Sweden are solidly placed within the anti-whaling block and they both joined the IWC when it became fashionable to do so at the end of the 1970s and early 1980's. Like so many of the other newly-recruited IWC members, they had previously not had any interest or involvement in the issue. It appears however, that more recently some nuances between the two can be perceived; while Finland is an unambiguous (yet very passive) protectionist country, Sweden is a more active player and seemingly more open to playing the role of a broker. The positions of Finland and Sweden are not too difficult to explain. This is a fairly simple issue for them just as it is for the large majority of the non-whaling nations of the IWC. As no material interests are involved, anti-whaling is a simple way to earn points in the international "green beauty contest". In the eyes of most observers it looks good internationally and, not least, gives the decision-makers some easy points from the domestic ENGO communities.[4] There are so many issues where it is difficult to be green, so why not grab the opportunities that exist? In my opinion this is a line of reasoning that *all* countries make use of when given the opportunity, not least Norway. Few countries are better at waving the green flag when no material interests are involved. As such, there is nothing wrong in using such opportunities for "symbolic policy". However, you have to be prepared for meeting the circumstance when *you* are the one with material stakes involved, and it may be painful to see that a large majority of "have-nots" vote you down.

At the other extreme of the continuum is Norway, the only country conducting commercial whaling - within the framework of the IWC. Iceland is put in brackets, as not being in quite the same category as Norway, neither conducting whaling nor being a member of the IWC. Nevertheless, there is little doubt that Iceland is positive to commercial whaling. In what may seem to be a difficult middle position is Denmark. On the one hand it is against commercial whaling, having a fairly green environmental profile to consider. On the other hand it has to consider the interests of Greenland and the Faroe Islands, which have very different attitudes towards the whaling issue, demonstrated by their memberships in NAMMCO. Although the situation is made somewhat easier by the fact that these two actors have sovereign jurisdiction over their living resources, it is neverthless Denmark that is the state representative in the IWC. However, as long as the IWC continues to endorse aboriginal whaling and not commercial whaling, Denmark can wave the green flag, and still be a whaling nation.

Although interests vary strongly, the Nordic countries have still found it useful to have deliberations on the issue. Thus, since the middle of the 1980s, the Nordic IWC Commissioners have had a separate meeting prior to the yearly IWC meetings. Moreover, in 1992 a report was commissioned by the Nordic Council to identify the positions of the Nordic countries on the whaling issue. My impression is that Sweden and Finland appear more sympathetic towards Iceland and Norway when the question is being discussed within the Nordic cooperative framework; language and attitudes seem to vary somewhat, depending on the setting. However, if there is a true shift in the Swedish position towards

---

[4] The whaling issue is *one* important reason for the varying strength of i.a. Greenpeace in Sweden and Norway; in Norway there are only a few thousand members while there are more than two hundred thousand Greenpeace members in Sweden.

the middleground, and this is not only linked to shift of personell, Sweden may have an important role to play in the future. What is needed in the IWC, if it is to get out of its current stalemate, is some countries that truly stand out as brokers between the antagonists.

The two most important countries in this setting are Iceland and Norway. Here I will start out by summarizing a few factual observations on the whaling policies of these two countries in recent years.

NORWAY
- conducted commercial whaling at the time of the adoption of the so-called moratorium;
- objected to the moratorium;
- continued a limited commercial catch until 1987;
- launched a scientific strategy in 1988, including scientific whaling, first very modest then gradually increasing catches;
- threats of economic sanctions from the US that never materialized (1986 - 1994). Certified 4 times;
- consumer boycotts organized by Greenpeace and other ENGOs (on and off between 1986 and 1994);
- resumed a limited commercial catch in 1993;
- commercial catch has increased gradually, especially since 1995;
- a member of both IWC and NAMMCO.

ICELAND
- conducted commercial whaling at the time of the adoption of the so-called moratorium - did not object to the moratorium;
- no commercial whale catch after 1985;
- launched a scientific strategy in 1986, gradually reducing scientific catch due to external threats;
- threats of economic sanctions from the US that never materialized, never certified;
- consumer boycotts organized by Greenpeace and NGOs 1988 - 89;
- no catch of whales after 1989;
- left the IWC in 1992;
- promoter in building up a possible alternative to the IWC,
- NAMMCO, established in 1992.

As we can see there are striking similarities between the two countries, but there are some very important differences as well.[5] With this as a point of departure, what kind of policies will these two countries pursue in the future, not least linked to the future development of NAMMCO?

NAMMCO: ACCOMPLISHMENTS AND FUTURE POTENTIALS

What has been accomplished?
How do we judge NAMMCO so far, what has been achieved along different dimensions? It is a fairly new organization and experience from other international organizations and regimes indicates that one cannot expect much of substance to happen within such a short time-span (Andresen & Wettestad, 1995). In a comparative perspective it seems to be quite successful, as quite a bit has been accomplished on the scientific side, and a lot of procedures i.a. linked to management and inspection are already in place.[6] Without doubt it has also had a positive effect regarding confidence-building and learning between the members. Thus, judged by these rather "soft" criteria it seems to have been fairly succecssful.

The question is what is going to happen next. Will NAMMCO be a "procedural success" but an empty shell from a management perspective, at least regarding whales under IWC jurisdiction? This in part depends on what the members of NAMMCO want it to be. If they want to stick to the present "scientific, knowledge-builiding and procedural" strategy and not interfere with substantial matters within IWC competence, things may be fairly easy and it will certainly be useful. However, things get more difficult if some of the members want to pursue a more ambitious strategy. As I am not a lawyer I will refrain from going into legal details regarding the relation between the Law of the Sea Convention, IWC and NAMMCO. I will stick to the policy side of the question, and it may well be the most decisive aspect in the final end.

The scope of membership
A critical question for the future of NAMMCO concerns the scope of membership. As it stands now, the parties *may* be too few to represent a viable management alternative as Norway and Iceland are the only two independent state-members. The membership of the Faroe Islands and Greenland certainly adds legitimacy and as such they are important, but

---

[5] In the forthcoming chapter from the IIASA project mentioned in footnote 2, I discuss the reasons behind these differences and similarities as well as the consequences of the different strategies adopted.
[6] For an overview of the status of what NAMMCO has achieved, see NAMMCO, *Annual Report 1996*, NAMMCO 1997.

it does not fundamentally alter the fact that there are only two state members. Before speculating on what the two key members want to use NAMMCO for, let us look at the possibilities for a broader state participation. There are several states attending NAMMCO meetings as observers.[7] Among these Russia and Japan are potentially new members.

One of the safest bets that can be made about Russia´s future policy is that it is *unpredictable*. The political situation is in turmoil; with cross-cutting cleavages between the various segments of the bureaucratic, political, business and military communities. One indication that *"rational politics"* can not be expected is reflected in the fact that the Russian Ministry of Environment was recently closed down - irrespective of the enormous environmental challenges facing the country. The severe difficulties that Norway as well as other western countries have faced in dealing with the various environmental challenges in Russia also show that Russia is no easy counterpart to deal with these days. Given this setting, the issue of whaling can hardly be expected to be given very high priority.

Nevertheless, let us speculate from a "rational perspective" on what Russia might do. Over the last few years Russia does not seem to have given very high priority to the IWC. Delegations have been very small and those sent do not appear to be high ranking or representing the more powerful segments in the Russian society. Moreover, Russia has not been given much support by the IWC majority. First there is the question of massive underreporting of catches by the previous Soviet Union. (Stoett, 1995). This also illustrates the internal strife on the issue in Russia. At the last IWC meeting in 1996, Russia was not allowed to take the aboriginal quota of 5 grey whales. Later on, Russia declared that it would disregard the IWC in this respect. It is my understanding that these whales may be far more important for these aboriginal people as a source of food than may be the case with other aboriginal whaling. In such a perspective, the '"green" arguments about conservation and protection of whales conveyed by the IWC majority cannot be expected to carry much weight in Russia. Conversely, it may be far more provocative to the relatively poor Russia than to the other (previous) and present whaling nations. In short, the above arguments indicate that IWC may not be considered very important to Russia. However, it should be recalled that apart from Norway, Russia is the only country that objected to the moratorium. Moreover, in contrast to most previous whaling countries Russia did *not* stop whaling due to pressure from the US and/or the environmental

---

[7] At the NAMMCO meeting in 1996 the following were present: Canada, Denmark, Japan, Namibia and Russia.

NGOs, but due to economic considerations. Thus, in the same manner as Norway, in legal terms, no-one can stop Russia should it want to conduct *commercial* whaling within the IWC.

Assuming that Russia wants to carry out some kind of whaling, will it do so as a small and "harassed" minority in the IWC or would it rather prefer to build a possible alternative to the IWC through full membership in NAMMCO or within some other alternative organization? It is my guess that the Russians have not decided on these questions yet, so I will refrain from further speculation. Just a small note in case Russia does become a NAMMCO member. As a point of departure such an important country certainly would broaden the NAMMCO basis and increase its legitimacy. However, with Russia on board, the attack on NAMMCO from the environmental movement might increase considerably. Although all previous Antarctic whaling nations should realize that they are marked by "the shadow of the past", this shadow may be particularly dark for Russia, considering the previous massive cheating that has been unravelled. In this perspective, a Russian membership may not have advantages only.

These last arguments do not apply to Canada. Canadian membership would increase the legitimacy of NAMMCO considerably. Canada left the IWC a long time ago and has considerable interests both as a sealing nation and as a small scale (aboriginal) whaling nation. It is also rumoured that Canada will in fact join NAMMCO before the next NAMMCO meeting in May this year (1997). The legal position of the aboriginals is strong in Canada. Considering the strong position of aboriginals in NAMMCO, a membership might seem a logical step. A key factor in the Canadian calculation on whether to join NAMMCO or not, is bound to deal with what effect it might have on the relationship to the USA. The Canadian - USA relations are currently quite strained, both on more general political issues (i.a. Cuba) as well as more specifically on the issue of marine mammals. The question is what effect these strained relations will have on the Canadian consideration about whether or not to join NAMMCO; should further escalations be avoided, or is there nothing to lose by membership? It may well be that Canada, a significant builder of international institutions in the North, would enjoy belonging to an organization dealing with northern affairs, that did not have the US as a member. So far it seems Canada has experienced the same as Norway has; there are threats and certifications but it does not come to sanctions; - would the same thing happen if Canada joined NAMMCO? I will get back to the US position towards NAMMCO in the final section.

Management body or 'bargaining chip'?

How many and which countries are going to join NAMMCO - if any - is certainly an important question for the future of NAMMCO; a new momentum may be created or a notion of stagnation may become apparent. However, *irrespective* of this question, what do the two main countries, Norway and Iceland want with NAMMCO? Does one or both of them really want it as a basis for managing whaling, or is it more of a "bargaining chip" in a broader game with other players?

To my knowledge, Iceland took the initiative to create what later on became NAMMCO. Why was this move made? It may be linked to the fact that Iceland was the first country to really take a serious beating within the IWC. While Norway was still a reluctant and cautious player, taking time to elaborate a strategy on the issue, Iceland was well underway with a scientific offensive in 1986. Iceland got much credit for for the scientific work in the IWC Scientific Commmittee, but politically not much was achieved. Against this backdrop it makes sense to start working for an alternative management regime; either as a bargaining chip to increase chances of getting your policies accepted, or to have it as a real fallback position in case of failure.

When it became clear to Iceland in 1991 that the rational scientific strategy would bear no fruits in the IWC, it may have seemed logical to leave the IWC and put all the apples in the NAMMCO basket.

Norwegian small-scale whalers as well as the Fishermen's Organization were happy with the Icelandic initiative. The Norwegian authorities also went along with the idea, partly as a means to pacify the strong opposition from these domestic actors towards what they saw as a rather defensive Norwegian position in the IWC at the time and partly to have more options available for future policies. However, with the benefit of hindsight we now know that the *main* Norwegian strategy was to start commercial whaling *within* the IWC. The same year as NAMMCO was created, Norway declared that commercial whaling would start within the framework of the IWC. Thus, as it turned out, for Norway NAMMCO has been essentially a bargaining chip. How important this card has been, especially in the bilateral negotiations with the USA, is not clear, but it has probably served a useful purpose, provided that the Norwegian threat of relying on NAMMCO if it did not get its way in the IWC was credible. This is not to say that NAMMCO has been and still is valued as a very useful instrument for the purposes that it is set up to serve. As it stands now, however, at least from a pragmatic point of view, I can see little reason for Norway to try to alter the de facto functioning of NAMMCO. Why try something new and uncertain when you basically have it your way in the present regime? I would assume that there is some kind of tacit understanding between Norway and the US that unofficially Norwegian commercial whaling will be accepted as long as it takes place within the IWC. I am not sure this understanding would be valid if Norwegian whaling were to take place within NAMMCO.

This is in accord with the fact that Norway sees itself as a true "internationalist", stressing the significance of law and order internationally. Being a small country this is generally in Norway´s interest. However poorly the IWC may be seen to function, as measured against what it was intended to be, or could have been, it is a fact that by the large majority of member states it is seen as the only *legitimate* international body for dealing with the whaling issue. These nations include the US, all the main EU states, most major western countries, including most Nordic countries. Conversely, I find it most probable that all these actors will view NAMMCO as *illegitimate* from a political perspective, irrespective of legal factors.

Although I believe this is the main calculation done by Norway until now, the picture is still not that simple. If the IWC majority continues to refute science as the basic management principle and continues to stall the implementation of procedures that may make commercial whaling possible, it may be difficult in the long run to continue spending time and resources on such an organization. In an "ideal world", where there were only whales to consider, I am quite sure that Norway would leave the IWC, but the reality is not as simple as that.

What about Iceland? The future policies of Iceland will probably to some extent be linked to the NAMMCO membership issue, and the kind of expectations and preferences the new potential members bring with them. Nevertheless, I think the basic question not mainly concerns the development of NAMMCO but *internal* Icelandic policy: *will Iceland resume commercial whaling?* My impression is that Iceland does not really know, or maybe does not want to make a decison? It is a fact that although the question has been extensively discussed for 5-6 years by Government(s) as well as Parliament in different settings, no decision has been taken.

There seems to be more internal disagreement on this issue in Iceland than there was in Norway. Clearly, actions by the US and parts of the NGO community are expected if whaling is resumed; actions that may divide the Icelandic society, depending upon their strength and duration. In any case, Iceland has to find out what it wants to do; re-enter the IWC, work to strengthen NAMMCO or simply declare that whaling is no longer a viable alternative in view of the present international climate. The *'wait and see strategy'* has now lasted so long that it seems high time to make some decisions.

SOME CONCLUDING OBSERVATIONS

The future of NAMMCO, the IWC, as well as increased whaling in general depends to a large extent on *public opinion* in western societies. Although the attitude towards whaling is not as negative as commonly perceived in anti-whaling countries like USA and UK (Freeman & Kellert, 1992), these countries probably have little to gain by changing their attitudes. Not least the NGO community will see to that. Nevertheless, public opinion tends to be cyclical, - the perception of marine mammals as special cretaures may also change; a few more perceived or real disatsers among livestock and other animals may be what it takes. It is also my impression that although the NGO community showing up at the annual IWC meetings shows no signs of decline, the issue is not quite as a 'hot' as it used to be. Moreover, an *alternative* NGO community, sympathetic to the whaling nations, parts of it highly skilled and with good understanding of the role of the media and politics, has succeeded in bringing more nuances into the picture. In short, it gets increasingly difficult both for politicians as well as for "green" NGOs to claim that whales *as such* are a threatened species, not the least due to the increasingly clear and consensual scientific message.

Although these are some important *trends*, they may be small comfort to those hoping that things will happen fast. In the short run, the key to the future of the whaling issue lies in the hands of the USA. It seems that its position in the IWC has for various reasons been reduced somewhat recently, but the US is still the most dominant and powerful player. It is my impression that the US strongly supports the IWC as the only legitimate basis for the management of whales; NAMMCO will not be tolerated as a competitor to the IWC. If this is the opinion of the USA, it is important that it uses its influence to make the IWC into an organization where careful and scientific management of whales will become feasible. If not, the whaling issue may continue to cause conflicts among otherwise like-minded nations. In a world where there are so many *real* problems relating both to the environment and the management of living resources, that should not be necessary. However, this may be a rather rational perspecive, and as politics are often irrational, the outcome is likely to be uncertain.

Acknowledgements:

I want to thank Alf Håkon Hoel, University of Tromsø, for useful comments on an earlier version of this paper.

REFERENCES

ANDRESEN, S., 1997, The making and implementation of whaling policies. Does participation make a difference? Forthcoming in Victor, D., Raustiala, C. and Skolnikoff, E., 1997, *The Implementation and Effectiveness of International Environmental Committments*, MIT PRESS, 1997.

ANDRESEN, S and Wettestad, J., 1995, International problem-solving effectiveness. The Oslo Project story so far., *International Environmental Affairs*, 7(2): 127-150.

FREEMAN M.R. & Kellert S., 1992 *Public Attitudes towards Whales, results of a six country study*, University of Alberta/University of Yale

Hvalfangstpolitikk 1992 - De Nordiske lands posisjoner i spørsmål om internasjonal forvaltning av hval, *Nordiske Seminar og Arbejdsrapporter*, 1992:584, Nordisk Ministerråd.

IVARSSON, J., 1994, *Science, Sanctions and Cetaceans, Iceland and the whaling issue*, Center for International Studies, University of Iceland.

HOLT, S., 1985, Whale mining, whale saving, *Marine Policy*, (3): 192-215

NAMMCO, *Annual Report 1996*, NAMMCO, 1997

STOETT, P, 1995, The International Whaling Commission: from traditional concern to an expanding agenda, *Environmental Politics*, 14 (1) 130-35.

Whales, the U.S. Pelly Amendment and international trade law

Ted L. McDorman *Associate Professor, Faculty of Law, University of Victoria, Victoria, B.C., Canada*

INTRODUCTION

The principal debating points about whales are: Should they ever be harvested? - the moral question. Is there sufficient scientific certainty to condone controlled harvesting? - the scientific question. Is whale harvesting illegal? - the legal question. None of these fundamental questions about whaling concern international trade law. It is the U.S. Pelly Amendment which connects the whale world with the international trade law world of the General Agreement on Tariffs and Trade (GATT) administered by the World Trade Organization (WTO).

Pursuant to the U.S. Pelly Amendment, countries that engage in whaling or trade whale products, irrespective of whether the whaling is consistent with the International Convention for the Regulation of Whaling or whether the country is a party to that treaty, may face trade sanctions against any of their products that enter the United States. While the United States has yet to impose a trade embargo against a country because of its whaling practices, the authority to do so is a continuing threat to states that permit their nationals to whale. Norway has been subjected to the Pelly process because of its 1990s decision to resume harvesting of minke whales, Japan because of its taking of whales pursuant to the research provisions of the International Convention for the Regulation of Whaling, and Canada has received U.S. attention because of the capture of two bowhead whales by the Nunavut in the Canadian Arctic.

The specific points to be examined in this contribution are: the operation of the Pelly Amendment; whether a trade embargo imposed under the Pelly Amendment would be consistent with U.S. obligations under the GATT; and whether the existence of the Pelly Amendment and the ever-present threat of trade sanctions is consistent with U.S. obligations under the GATT.

THE PELLY AMENDMENT

The threshold step in the possible use of the Pelly Amendment to impose a trade embargo against a whaling country is certification. Only when the U.S. President receives a letter of certification may a trade embargo be imposed.

The Pelly Amendment establishes two avenues for certification. A third whaling-specific avenue for certification exists through the Packwood-Magnuson Amendment. However, a certification under the Packwood-Magnuson Amendment is deemed to be a certification under the Pelly Amendment and the trade embargo provisions of the Pelly Amendment are invoked, hence the Packwood-Magnuson Amendment is usually grouped with the Pelly Amendment process. The two amendments require a letter of certification to be sent to the U.S. President where:

i) the Secretary of Commerce determines that foreign fishing is *diminishing the effectiveness* of an international fishery conservation program;
ii) the Secretary of Commerce or Interior finds that foreign nationals are capturing or engaging in trade which *diminishes the effectiveness* of any international program for endangered or threatened species; or
iii) the Secretary of Commerce concludes that foreign fishing or trading *diminishes the effectiveness* of the International Convention for the Regulation of Whaling. (emphasis added)

The second of the three avenues creates the possibility of certification where a country is trading in endangered species and thus diminishing the effectiveness of the Convention on International Trade in Endangered Species (CITES). The first and third avenues create the possibility of certification where a country is harvesting or trading whale that is diminishing the effectiveness of the International Convention for the Regulation of Whaling. The first avenue captures this because an international fishery conservation program specifically includes the International Convention for the Regulation of Whaling.

The diminishing the effectiveness criterion at the heart of the Pelly Amendment certification decision is imprecise since it is not elaborated upon or defined in the legislation. In order to encourage stricter compliance with international fishery agreements, CITES and the International Convention for the Regulation of Whaling, it is argued by some that a broad view of what activities diminish the effectiveness of these arrangements is necessary. The international concern is whether the United States should have the role of enforcing international agreements, particularly where the United States establishes the standard that must be met by other states and unilaterally judges whether there has been a breach of that standard.

Upon receipt of a certification made under either the Pelly or Packwood-Magnuson Amendments, the U.S. President may impose a trade embargo against the certified country. Prior to 1992, a Pelly embargo was to be restricted to either fish or wildlife products from the certified country. Following 1992 amendments, a trade embargo can be placed against *any* products from the certified country. Thus, the trade threat from the Pelly Amendment has been increased.

However, the U.S. President has a wide discretion regarding the imposition of an embargo. Thus far, the U.S. President has not imposed an embargo on a whaling state certified under either the Pelly Amendment or the Packwood-Magnuson Amendment. Advocates for employment of the Pelly and Packwood-Magnuson Amendments see them as a way to use the economic power of the United States to encourage or force renegade countries to adhere to the whale quotas and moratorium of the International Whaling Commission. Concerns that the United States is imposing *its* standards of behaviour on foreign actors operating outside the United States and the unilateral nature of the Pelly process are swept aside in the moral fervor over whales.

TRADE LAW RELEVANT TO A PELLY AMENDMENT EMBARGO[1]

As the principal architect of both the GATT, the basic treaty of international trade law, and the WTO, the organization which administers the GATT, the United States is constrained by the rules and norms of both the GATT and the WTO. The fundamental constraint of WTO membership is that, except in compliance with the GATT, WTO members cannot unilaterally determine with whom and how they will trade with other WTO members.

The GATT Rule on Trade Embargoes

Regarding trade embargoes, the fundamental rule of the GATT is Article XI(1), which prohibits WTO-members from imposing quantitative restrictions (quotas) or quantitative prohibitions (embargoes) on imports or exports. Such restrictions are seen as inconsistent with the free flow of goods, the operation of comparative advantage, and an efficient allocation of world resources, all of which are objectives of the GATT rules.

The no-trade embargo rule of Article XI can be avoided where the GATT is superseded by other international obligations and therefore not applicable, or, even where the GATT is applicable, one of the numerous exceptions in the GATT applies.

The Non-Application of GATT Article XI

There are two situations where GATT Article XI would not be applicable to a trade embargo: (a) where the trade embargo was required by another multilateral international treaty, and (b) where the trade embargo was a countermeasure or reprisal for breach of a treaty.

Trade embargoes will be neither GATT illegal, nor perceived as unacceptable, where a multilateral international treaty explicitly permits the employment of such a trade measure. The prime example is CITES which explicitly requires its members to utilize embargoes against the import and export of endangered species. A CITES trade embargo is a *prima facie* inconsistency with the GATT. CITES would override GATT obligations where the states involved are members of both CITES and GATT. It is a delicate legal and political question whether CITES obligations would override GATT obligations if the states involved in a dispute were members of GATT but not CITES.[2] (See below regarding Iceland). It is important to note that CITES does not permit trade embargoes to be employed

---

[1] This section draws from T.L. McDorman, "Protecting International Marine Living Resources with Trade Embargoes: GATT and International Reaction to U.S. Practices" in G. Blichfeldt, ed. *Additional Essays on Whales and Man* (Reine i Lofoten, Norway: High North Alliance, 1995), at pp. 21-28.
[2] See: T.L., McDorman "The 1991 U.S.-Mexico GATT Panel Report on Tuna and Dolphin: Implications For

because a country is not appropriately protecting endangered species.

Trade embargoes connected to protection of the marine living resources would not be subject to GATT scrutiny where a trade embargo was imposed as a countermeasure or reprisal for the *breach* of an international environmental treaty or marine living resource treaty. Such a trade embargo would be given priority over potentially conflicting GATT obligations since trade measures are a legitimate way to deal with treaty breaches, although reprisals generally are used sparingly and involve, amongst other conditions, proportionality. As a matter of international law, only those countries explicitly bound by an international environmental or ocean resource protection treaty can utilize or be subject to enforcement embargoes of this type.

The Exceptions to Article XI

The two exceptions in the GATT that are seen as having the potential to justify trade embargoes for marine living resources conservation purposes are Article XX(b) and (g). Article XX(b) allows governments to impose trade measures that are "necessary to protect human, animal or plant life or health". Article XX(g) permits governments to utilize trade measures, such as import embargoes, where the embargoes are "relating to the conservation of exhaustible natural resources if such measures are made effective in conjunction with restrictions on domestic production or consumption". Both Article XX(b) and (g) are qualified by the requirements that any trade measures taken must not constitute arbitrary or unjustified discrimination and must not be a disguised restriction on trade.

Relying on the GATT exceptions, there is no question that a state can utilize trade embargoes to protect marine living resources *subject to* or *within* their own jurisdiction. More generally, the GATT imposes few constraints on policies and practices designed to protect a state's domestic environment. The only qualification is that a trade measure must have a bona fide environmental purpose and not be a disguised protectionist measure.

The difficulty in application of the GATT exceptions arises where the marine living resource being protected is not within the jurisdiction of the state imposing the embargo and the actors are foreign nationals operating pursuant to foreign law. The wording of Article XX(b) or (g) is not restricted by geographic location or nationality, thus could be interpreted to accommodate an embargo designed to protect transnational marine

---

Trade and Environment Conflicts" (1992), 17 *North Carolina Journal of International Law and Commerce* 461, at pp. 484-485.

living resources being harvested by foreign nationals.

The 1994 *U.S. - E.U. Tuna-Dolphin* GATT dispute settlement panel examined this issue respecting a trade embargo imposed under the U.S. Marine Mammal Protection Act triggered by foreigners harvesting tuna in waters beyond U.S. jurisdiction in a manner inconsistent with U.S.-created standards for dolphin protection.[3] The panel concluded that the U.S. tuna embargo was inconsistent with Article XI and that neither Article XX(b) or (g) exceptions were applicable. The Panel concluded that Article XX(b) or (g) were not applicable since the embargo was unrelated to conservation of domestic resources and, more importantly, was designed to force other countries to adopt laws and practices compatible with the unilaterally-determined U.S. standard.

APPLYING INTERNATIONAL TRADE LAW TO THE PELLY AMENDMENT

The Threat of an Embargo

The certification process of the Pelly and Packwood-Magnuson Amendments and the requirement of the U.S. President to consider the imposition of an embargo creates an economic threat against states. Does the GATT prohibit economic threats such that the Pelly and Packwood-Magnuson Amendments can be challenged under the GATT, found inconsistent with GATT rules and the United States be required to remove the offending legislation?

The 1991 *U.S. - Mexico Tuna-Dolphin* GATT dispute settlement panel looked directly at this issue and found that "because the Pelly Amendment did not require trade measures to be taken, ...(it)... was not inconsistent with the General Agreement".[4] The panel determined that legislation merely creating the possibility of imposing a trade embargo was not inconsistent with the GATT. This finding, consistent with previous determinations, ensures a broad scope for domestic law free from the interference of the GATT. Thus, the Pelly and Packwood-Magnuson Amendments themselves are immune from GATT attack, although any embargo imposed could be subject to GATT questioning.

The Case of Norway

As discussed in section three, not every trade embargo runs afoul of the

---

[3] *United States - Restrictions on Imports of Tuna*, Report of the GATT Panel, May 20, 1994, reprinted in (1994), 33 *International Legal Materials* 842.

[4] *United States - Restrictions on Imports of Tuna*, Report of the GATT Panel, August 16, 1991, at. para. 5.21, reprinted in (1991), 30 *International Legal Materials* 1594.

GATT rules and, therefore, not every embargo that might be imposed pursuant to the Pelly Amendment would be inconsistent with the GATT. The specific circumstances of an embargo would need to be assessed - no blanket statement can be made. However, it is possible to determine that a Pelly Amendment embargo arising from recent Norwegian whaling would *not* be GATT consistent.

First, if a Pelly Amendment embargo was imposed against a country, a party to the IWC, that was in breach of the Whaling Commission rules, then the trade embargo might be considered a reprisal or countermeasure. Such a reprisal or countermeasure would override the rules of the GATT. Only states that are parties to the IWC, however, can be in breach of the treaty and thus be subjected to a reprisal. Norway, unhappy with the continuation of the whaling moratorium, but consistent with its law-abiding manner, elected to utilize the provisions of the Whaling Treaty to "opt out"of the moratorium resolution. Norway's decision to renew commercial whaling, therefore, is unquestionably legal under the International Convention for the Regulation of Whaling.
Norway is using the whale products internally and is not engaged in trade of whale products. Moreover, the status of the whales being captured by Norway as endangered species is open to debate. The result is that Norway is not attracting attention under CITES. Thus, there is no question of whether CITES overrides the GATT.

Prima facie a U.S. Pelly Amendment embargo would be inconsistent with GATT Article XI(1) and thus would need to fit within Article XX(b) or (g). Following the *1994 U.S.-E.U. Tuna-Dolphin Panel* decision, the critical consideration is whether the embargo is related to the protection of whales in U.S. waters or whether the embargo is designed to force the other state to follow the quota or moratorium of the IWC, determined by the United States to be the applicable standards. If the former, Article XX(b) and (g) may be successfully utilized and the trade embargo determined to be GATT consistent. If the latter, the trade embargo would be inconsistent with the GATT. In the case of Norwegian whaling, it is clear that any U.S. embargo would be designed to deal with production questions in Norway, and, unquestionably, the situation would be one of the United States imposing or forcing its whale standards on another state.

The Case of Canada
Like Norway, Canada is not a renowned environmental delinquent. The

decision to permit limited harvesting of bowhead whales was based on the social and cultural significance of the whales to the Nunavut. Moreover, Canada is no longer a party to the IWC and thus cannot be in breach of the IWC. The GATT exceptions would not support a Pelly Amendment embargo since the embargo is unrelated to protection of whales in the United States and is designed to force Canadian compliance with the whaling standard the United States deems appropriate.

The Case of Iceland
Like Canada, Iceland is no longer a party to the International Convention for the Regulation of Whaling. A resumption of whaling by Iceland, therefore, cannot be a breach of the IWC and a U.S. trade embargo could not be justified as a reprisal.

A trade embargo against Iceland premised solely upon a resumption of whale harvesting would be inconsistent with GATT for the same reasons such an embargo against Norway or Canada would be GATT-illegal. The exceptions to the GATT "no embargo"rule do not apply where the embargo is designed to deal with production issues (i.e. whaling) *in* Iceland, which do not affect the United States, and where the principal purpose of the trade embargo is to force compliance with U.S. - determined standards. Unlike the Norwegian and Canadian situations, however, Iceland is contemplating trade in whale products which may lead to trade embargoes premised on CITES concerns. Iceland, however, is not a member of CITES. If the whale sought to be exported by Iceland was not subject to CITES (not listed in Appendix I), then a trade embargo against either whale products or other Icelandic products would be GATT-illegal. If, however, the exported whale was subject to CITES (listed in Appendix I), then an import embargo against the whale products or other Icelandic products *may* be GATT legal.

As a technical, formal legal issue, since Iceland is not a party to CITES, the GATT rules would apply between Iceland and the United States rather than CITES and any embargo against Iceland based upon CITES, or a "breach"of CITES, would be GATT-illegal. However, if a U.S. embargo were restricted to whale products from Iceland, in other words, the United States was adhering to its CITES obligation, it is highly unlikely that a WTO dispute settlement panel would decide that the U.S. embargo was GATT-inconsistent. If the U.S. embargo was against non-whale Icelandic products and imposed because of Icelandic whale exports to a third country, a WTO dispute settlement panel would be confronted with the difficult issue of the United States attempting to use its economic prowess to enforce CITES, a widely-accepted regime, against a non-member of CITES. While this scenario is similar to the tuna-dolphin cases, a key difference is the broad international acceptance of CITES. The sensitivity of the WTO to attack as being anti-environmentalist also plays a role. The technical law suggests a favourable response to Iceland, the U.S. embargo being illegal, however, a WTO dispute settlement panel would be very *hesitant* to undermine the principles of CITES and impose the GATT rules.

CONCLUSION

The Pelly and Packwood-Magnuson Amendments permits the United States to impose a trade embargo against any products from a state certified as harvesting or trading whales in a manner which diminishes the effectiveness of either the International Convention for the Regulation of Whaling or the Convention on International Trade in Endangered Species (CITES). The "diminishing the effectiveness" criterion is flexible enough to allow U.S. embargoes irrespective of whether the impugned state is a party to either treaty or has operated illegally under either treaty. Thus far, however, the U.S. President, relying upon the discretion in the legislation, has not imposed an embargo against a whaling state certified under the legislation.

The issues addressed in this contribution concern the legality under the rules of the GATT of Pelly Amendment action. First, the threat of an embargo is not GATT illegal. Second, no blanket answer can be given whether a Pelly embargo would be inconsistent with the GATT. Each circumstance needs to be examined. In the recent cases of whaling by Norway and Canada, if an embargo had been employed by the United States such an embargo would have been illegal under the GATT. A similar result would be the case if Iceland were to resume whaling, provided Iceland did not trade in whale products. Despite not being a member of CITES, a U.S. embargo imposed because of Icelandic trade in whale products would create a direct confrontation between the law and politics of CITES and the GATT. It would be a legally-challenging and politically-charged confrontation which makes prediction difficult.

CITES and international trade in whale products

Jaques Berney *Adviser, CITES Secretariat, 15, Chemin des Anémones, P.B. 456, CH-1219 CHÁTELALINE - Genéve, Switzerland   berneyj@unep.ch*

WHAT IS CITES?

CITES, the Convention on International Trade in Endangered Species of Wild Fauna and Flora, is an international legal instrument that was signed in Washington D.C., on March 3rd 1973. It entered into force on July 1st 1975. It has now a membership of 135 States, the Parties, including all those surrounding the North Atlantic, except Iceland and Ireland. Although the Faeroe Islands are covered by the ratification of Denmark, the Convention is still not implemented in that territory due to the absence of appropriate legislation.

The aim of the Convention is to ensure the co-operation of the Parties to prevent international trade in specimens of wild animals and plants from being detrimental to their survival.

BASIC PRINCIPLES

CITES works by subjecting trade in specimens of species of wild fauna and flora included in three appendices to certain controls, through a system of permits and certificates granted by Management Authorities, on the advice of Scientific Authorities. Each Party must designate at least one Management Authority and one Scientific Authority.

In the context of CITES, "trade" means import, export, re-export and introduction from the sea. "Introduction from the sea", this is of significance for whales, means transportation into a State of specimens of any species which were taken in the marine environment not under the jurisdiction of any State. CITES has nothing to say about sale or movement of specimens within a country, including about the landing of specimens taken in the national waters of any country with sea coasts, or about the way the species are managed, used and conserved, and even killed. This is left entirely to national legislation.

The species covered by CITES are listed in three appendices, according to the degree of protection they need. Appendices I and II are revised at the regular meetings of the Conference of the Parties, which take place every two and a half years. The next meeting will be held in Harare, Zimbabwe, in June this year (1997). In case of urgency, Appendices I and II may be amended between regular meetings through a postal procedure rarely used. Appendix III may be revised at any time, but a Resolution of the Conference of the Parties recommends that this be done at the time of regular meetings.

> **Appendix I** includes species threatened with extinction. Trade in specimens of these species is permitted only in exceptional circumstances. Import permits and export permits or re-export certificates are required. In case of introduction from the sea, a certificate must be issued by the Management Authority of the country of introduction under strict conditions.

> **Appendix II** includes species not necessarily threatened with extinction, but for which the trade must be controlled in order to avoid utilization incompatible with their survival. It includes also species which must be subject to regulation in order that the trade in other listed species is brought under effective control. Export permits or re-export certificates only are required. In case of introduction from the sea, a certificate must also be issued by the Management Authority of the country of introduction under specified conditions but less strict than for Appendix-I species.

> **Appendix III** contains species that are protected in at least one country, which has asked other CITES Parties for assistance in controlling the trade in their specimens. Export permits are required

from the countries that have required the listing while certificates of origin are required from the other countries. By definition, introduction from the sea does not exist for Appendix-III species.

CRITERIA FOR AMENDMENT OF APPENDICES I AND II

When CITES was adopted in 1973 with the original Appendices I and II, these lists were adopted on the basis of lists prepared by IUCN-the World Conservation Union and of proposals submitted by participants to the plenipotentiary conference held in Washington D.C. While we can expect that the IUCN lists were based on certain criteria, it is obvious that a number of species were added without scientific justifications. Apart from the information provided above, the Convention does not provide for listing criteria.

At its first meeting (Berne, 1976), the Conference of the Parties adopted criteria for the inclusion of species in Appendices I and II, for the transfer of species from one appendix to the other and for the deletion from the appendices. It must be recognized, and it was, that under the Berne criteria the transfer of a species from Appendix I to Appendix II, or the deletion of a species from these appendices, was almost impossible, unless the species was listed under the criteria for addition. But none of the species included in Washington or in Berne were listed following the Berne criteria. Therefore, the Parties adopted special criteria to permit "downlisting" without meeting the Berne criteria.

At the eighth meeting of the Conference of the Parties (Kyoto, 1992), a Resolution was adopted to direct the CITES Standing Committee to undertake, with the assistance of the Secretariat, a revision of the criteria for consideration at the ninth meeting. The expertise of IUCN and other organizations and individuals was to be used for this process. A draft resolution was considered at the ninth meeting (Fort Lauderdale, 1994) and Resolution Conf. 9.24 was adopted by consensus, after long and difficult discussions.

The Resolution contains the criteria that have to be met to amend Appendices I and II in one way or another. There are biological and trade criteria. The Resolution refers also to special cases such as the listing of higher taxa than species and the listing of populations of the same species in various appendices (split-listing), and it includes precautionary measures to prevent decisions that could be detrimental to the conservation of the species concerned.

The new criteria will be used at the tenth meeting in Harare for the first time. It must be said however, that how good the criteria might be, this will have to be judged in the future, nothing would prevent the Conference of the Parties either to adopt amendments although the species concerned does not meet the criteria or, on the contrary, to reject amendments although the species concerned obviously meet the criteria. Politics, emotions and other non-scientific elements are not absent of CITES discussions and decisions, as has been demonstrated many times.

It is worthwhile to note also that if the criteria to amend the appendices have changed since the minke whale was included in Appendix I, resolutions adopted by the Conference of the Parties in relation to whale and whaling issues have not been significantly amended or repealed.

SPECIFIC RESERVATIONS

The provisions of the Convention are not subject to general reservations. However, specific reservations may be entered with regard to species listed in the appendices.

When a State is ratifying the Convention or acceding to it, it may, on depositing its relevant instrument, enter reservations with regard to species listed in Appendices I, II and III. However, it may not do so at a later stage, except for Appendix-III species.

When amendments to Appendices I and II are adopted by the Conference of the Parties, these amendments enter into force after a period of 90 days. During that period, and only during it, any Party may enter a reservation with regard to any of these amendments.

Until a Party withdraws its reservation it is treated as a State not party to the Convention with respect to trade in specimens of the species concerned.

The Conference of the Parties has adopted a resolution, Resolution Conf.

4.25 on the Effects of Reservations, which recommends that a Party having entered a reservation with regard to an Appendix-I species consider it as listed in Appendix II. Therefore, that Party should issue CITES trade documents and submit reports on the trade in that species.

It is important to note that if a Party with a reservation may trade freely in specimens of the species concerned, even if it is listed in Appendix I, the other Parties may not authorize such trade unless they have entered the same reservation or the trade is conducted in accordance with the relevant provisions of the Convention.

Regarding whale species, Japan has entered reservations with regard to the Baird's beaked whale *Berardius bairdii*, the sperm whale *Physeter catodon*, the minke whale *Balaenoptera acutorostrata*, except the West Greenland stock, the sei whale *Balaenoptera borealis*, except stocks (A) inthe North Pacific and (B) in the area from 0 degrees longitude to 70 degrees east longitude, from the equator to the Antarctic Continent, the Bryde's whale *Balaenoptera edeni* and the fin whale *Balaenoptera physalus* stocks (A) in the North Atlantic off Iceland, (B) in the North Atlantic off Newfoundland and (C) in the area from 40 degrees south latitude to the Antarctic Continent, from 120 degrees west longitude to 60 degrees west longitude. Norway has entered reservations with regard to the sperm whale *Physeter catodon*, the minke whale *Balaenoptera acutorostrata*, except the West Greenland stock, the sei whale *Balaenoptera borealis*, except stocks (A) in the North Pacific and (B) in the area from 0 degrees longitude to 70 degrees east longitude, from the equator to the Antarctic Continent, and the fin whale *Balaenoptera physalus*. Peru has entered reservations with regard to the minke whale *Balaenoptera acutorostrata*, except the West Greenland stock, the Bryde's whale *Balaenoptera edeni* and the pygmy right whale *Caperea marginata*. Finally, Saint Vincent and the Grenadines has entered a reservation with regard to the humpback whale *Megaptera novaeangliae*. All the species and stocks concerned are included in Appendix I.

Consequently, under CITES, nothing would prevent Japan and Norway, for example, to introduce from the sea and/or to trade between them in specimens of sperm whale, of minke whale and of some stocks of sei whale as both countries are considered as not party to CITES with regard to such introduction and trade.

SPECIAL PROVISIONS

In addition to exceptions and special provisions included in Article VII of the Convention, e.g. concerning pre-Convention specimens, personal effects and captive breeding, that are not really relevant to the purpose of this conference and on which I will therefore not insist, special provisions are made in CITES regarding other treaties, conventions or international agreements. These are very relevant to whales and are the following.

Article XIV on "Effects on Domestic Legislation and International Conventions" states in its paragraph 4 that "A State party to the present Convention, which is also a party to any other treaty, convention or international agreement which is in force at the time of the coming into force of the present Convention and under the provisions of which protection is afforded to marine species included in Appendix II, shall be relieved of the obligations imposed on it under the provisions of the present Convention with respect to trade in specimens of species included in Appendix II that are taken by ships registered in that State and in accordance with the provisions of such other treaty, convention or international agreement."

Paragraph 5 of the same Article states that "Notwithstanding the provisions of Articles III, IV and V [the reference to Article III and V is obviously an error in this paragraph because it is not relevant to trade in Appendices-I and -III species], any export of a specimen taken in accordance with paragraph 4 of this Article shall only require a certificate from a Management Authority of the State of introduction to the effect that the specimen was taken in accordance with the provisions of the other treaty, convention or international agreement in question."

Finally, paragraph 6 indicates that "Nothing in the present Convention shall prejudice the codification and development of the law of the sea by the United Nations Conference on the Law of the Sea convened pursuant to Resolution 2750 C (XXV) of the General Assembly of the United Nations nor the present or future claims and legal views of any State concerning the law of the sea and the nature and extent of coastal and flag State jurisdiction."

Regarding whales, if they are listed in Appendix II, the above-mentioned provisions clearly indicate that the role of CITES is very limited for those

States that are bound to the provisions of the International Convention on the Regulation of Whaling. However, all great whales, except the West Greenland stock of the minke whale, are included in CITES Appendix I at present.

CITES AND THE MINKE WHALE

The minke whale *Balaenoptera acutorostrata* was included in CITES Appendix II at the second meeting of the Conference of the Parties (San José, 1979) when all cetaceans not yet included in CITES appendices were included in that appendix. It is at the fourth meeting (Gaborone, 1983) that the minke whale, except the West Greenland stock, was transferred to Appendix I but the entry into force of this amendment, instead of normally taking place ninety days after the meeting, was postponed by the Conference until 1 January 1986, to coincide with the entry into force of the IWC moratorium on commercial whaling.
The proposal for the transfer, submitted by Seychelles, concerned in fact all species of the order Cetacea the catches of which were regulated by the IWC. It is of course not possible to report here details of the arguments of the proponent. The following, however, would appear of interest. The proposal stated: "There is no positive evidence that any putative population of minke whales is endangered in terms of the literal interpretation of the Berne criteria for CITES. On the other hand, given the total absence of scientific assessments for any minke whale stock we can not reasonably assume that they are less endangered than some of the stocks of other baleen whale species that now enjoy full protection by IWC and by inclusion in CITES Appendix I." At the meeting itself, the delegation of Seychelles decided, as allowed within CITES, to amend its proposal to exclude from the transfer to Appendix I the West Greenland stock, because it was the only population from which there was a take to meet aboriginal needs for local consumption, and the intention of the IWC was certainly that the decision regarding a zero commercial catch limit did not preclude such taking.

Some Parties expressed their opposition to the proposal already before the meeting and also during it, and the CITES Secretariat recommended that the proposal be rejected because it neither met the Berne criteria nor the provisions of the Convention. A CITES committee set up to review the species listed in CITES appendices had concluded earlier that no changes be made for cetaceans. After intense discussions, the Conference of the Parties voted upon the proposal, which was adopted as amended by 29 votes in favour, five votes against and 23 abstentions.

THE NORWEGIAN PROPOSAL

The Government of Norway has submitted a proposal to transfer the Northeast Atlantic and North Atlantic Central stocks of minke whale from Appendix I to Appendix II for consideration at the tenth meeting of the Conference of the Parties. Two questions have been raised by the High North Alliance in their letter to me concerning this presentation. What criteria will be decisive regarding the proposal and what will be the arguments in favour or against it?

The criteria to be used are those provided by Resolution Conf. 9.24. In my opinion, it is obvious that the minke whale is or may be affected by trade and it is known, inferred or projected that unless the trade is regulated the stocks in question will meet at least one of the criteria for inclusion in Appendix I, sooner or later. Therefore, the minke whale, or the stocks in question, should be listed at least in Appendix II. To decide whether they should be listed in Appendix I, the criteria for inclusion in that appendix must be considered. In summary, they are the following:

- is the wild population (stock) small, and characterized by at least one other factor amongst five?
- has the wild population a restricted area of distribution and is characterized by at least one other factor amongst four?
- has a decline in the number of individuals in the wild been observed or is inferred or projected?
- is the status of the species such that if it is not included in Appendix I it is likely to satisfy one or more of the above criteria?

I am not here to answer to these questions, and since the Secretariat has not yet adopted its formal position, I will not make any further comments on this although I have my own idea.

Regarding the arguments in favour or against the proposal, I believe that those in favour are well known to you as they are contained in the proposal itself. For those against, it is possible to refer to the discussion held in Fort Lauderdale on a similar proposal already submitted by Norway.

We can refer also to the comments sent to IWC by some of its members in response to the communication of the proposal.

- The current listing in CITES appendices is in recognition of the IWC zero quota for commercial whaling;
- the IWC moratorium remains in place and any move to facilitate trade would undermine it;
- the IWC accepted an abundance estimate for the Northeast Atlantic stock but agreed that further analyses should be carried out;
- no estimate for the Central North Atlantic stock has been approved;
- Norway has proceeded with a commercial catch despite the IWC resolutions;
- it is inappropriate to consider a transfer in advance of any IWC decision on the Revised Management Scheme;
- under precautionary measures in Resolution Conf. 9.24, the views of the IWC should be taken into account; there has been no IWC decision that would support the proposal;
- CITES Resolution Conf. 2.9, which recommends that the Parties do not issue any permit or certificate for any specimen of a species or stock protected by the IWC, is still in effect;
- the species or stocks covered by the IWC moratorium should remain in CITES Appendix I;
- in 1978, the IWC passed a resolution requesting that CITES take all possible measures to support the IWC ban on commercial whaling; the CITES Parties adopted then Resolution Conf. 2.9 and, in 1983, decided to amend the appendices accordingly;
- there is at present no International Observer Scheme in place to vet catches;
- how to distinguish in trade products from these stocks and those from other stocks?
- a transfer to Appendix II would run counter the IWC resolutions against Norway's setting of quotas on commercial whaling.

To these arguments, which, to a large extent, derive from discussions within and decisions from the IWC, and on which I will not comment, it is of course necessary to add the emotional views expressed by animal welfare and animal rights groups. They are certainly not negligible as they also influence the decisions of certain governments.

CITES TRADE-CONTROL MEASURES

I have been asked whether CITES may impose control measures to avoid legal trade hiding illegal trade. The answer is yes and no. The fact that species may be included in Appendix II because they look like other listed species is an example of such imposed control measures. This serves to prevent that an endangered or threatened species be traded under the name of a common species similar to it. In addition, all control measures established under the Convention and enforced by the Parties have as objective to prevent illegal trade, including that hidden by legal trade.

Through Resolutions, the Conference of the Parties has adopted a number of additional trade-control measures but they are not imposed as the Resolutions are not binding for the Parties, unless the provisions of these Resolutions are included in the national legislation of the Parties themselves. Amongst others, I can mention the control of shipments in transit through the territory of a Party, the marking of specimens from certain categories of species, strict rules concerning the issuance of retrospective permits and certificates, the establishment of export quotas (quotas may also be imposed if they are included in the appendices trough specific annotations). In certain circumstances, it is also recommended to the Parties to suspend the trade in CITES specimens with a particular Party, in general or regarding designated species. Such suspensions are usually recommended by the Standing Committee.

The self imposition of additional trade controls by a Party asking for the transfer of a species from Appendix I to Appendix II might of course help the adoption of such a proposal. For example, several proposals submitted for consideration at the tenth meeting include such measures, in particular those concerning the African elephant or the vicuna. In the case of whales, measures concerning the identification and the marking of the products in trade, as well as controls on the markets might be a prerequisite for any proposal to be adopted.

RELATIONS BETWEEN THE IWC AND CITES

Although the IWC is not mentioned in the text of the Convention, Article XIV, as indicated previously under Special Provisions, does actually refer to the IWC, as far as Appendix-II species are concerned. In addition, for marine species, Article XV of the Convention directs the Secretariat to "consult inter-governmental bodies having a function in relation to those species when a proposal to amend Appendix I or II is submitted". This obviously refers also to the IWC and has been so interpreted by the Secretariat.

The IWC Secretariat is always invited to participate as an observer to meetings of the Conference of the Parties to CITES and, in reciprocity, CITES Secretariat has obtained the observer status, and adviser status for trade matters, at meetings of the IWC and of its Scientific Committee.

The Conference of the Parties to CITES has adopted several Resolutions related to the IWC or to whales:

- Resolution Conf. 2.7 (Rev.), adopted in San José in 1979 and revised in Fort Lauderdale in 1994, which recommends that Parties not member of the IWC adhere to it;
- Resolution Conf. 2.8, adopted in San José in 1979, which recommends that Parties use their best endeavours to apply their responsibilities under the Convention in relation to cetaceans;
- Resolution Conf. 2.9, adopted in San José in 1979, which recommends that Parties agree not to issue any import or export permit, or certificate for introduction from the sea for primarily commercial purposes for any specimen of a species or stock protected from commercial whaling by the IWC; and requests that the Secretariat circulate to the Parties a list of species or stocks so protected;
- Resolution Conf. 3.13, adopted in New Delhi in 1981, which recommends that Parties pay particular attention to the documentation requirements for specimens of cetaceans under Articles IV and XIV; and that Parties give urgent consideration to Resolution Conf. 2.7;
- Resolution Conf. 9.12, adopted in Fort Lauderdale in 1994, concerning illegal trade in whale meet; and
- Resolution Conf. 9.24, adopted in Fort Lauderdale in 1994, concerning the new criteria.

At the ninth meeting in Fort Lauderdale, the Secretariat, within the process of consolidation of Resolutions, suggested to repeal Resolutions Conf. 2.7, Conf 2.8 and Conf. 3.13, considering them as redundant or out of date. This was not accepted by the Parties and these Resolutions are still in force, although Resolution Conf. 2.7 was amended.

At the tenth meeting, the Secretariat will submit again a consolidated draft resolution, which will include the provisions of Resolutions Conf. 2.7 (Rev.), Conf. 2.9 and Conf. 9.12 and repeal all the Resolutions mentioned above, except of course Resolution Conf. 9.24 .

The most significant Resolution, in practical terms, is Resolution Conf. 2.9, although it is in fact redundant for as long as the species protected by the IWC are included in Appendix I. Its effect would be, if a whale species or stock is transferred to Appendix II, to prevent commercial trade in that species or stock by the Parties implementing the Resolution, although, if this Party is also a member of the IWC, it will be relieved of CITES obligations in accordance with Article XIV, paragraph 4, mentioned earlier.

For the Secretariat, the Resolutions of the Conference of the Parties are binding and, therefore, it will not be in a position to make any recommendation contrary to a Resolution. Does that mean that the Secretariat must recommend to the Parties to reject any proposal concerning the minke whale? My answer, not that of the Secretariat as far as I am aware, will be no, since Resolution Conf. 2.9 does not request such a recommendation. However, if the Secretariat recommends the acceptance of a proposal on the minke whale on the basis of the criteria provided by Resolution Conf. 9.24, it will have to recommend also that the transfer be associated with a zero export quota to be maintained until either Resolution Conf. 2.9 has been repealed or amended, or the IWC has lifted the moratorium or decided to lift the protection granted to the stock in question.

Accordingly, CITES, under the text of the Convention and within the limits described in Article XIV, is not committed to act according to the political considerations in the IWC and it may act solely on its own assessment of the available scientific data. Under the Resolutions into force however, the Conference of the Parties has agreed not to do so and to follow the position of the IWC. This is of course reversible but would need a formal decision of the Conference of the Parties and you may wish to know that such a decision will require a two thirds majority vote as for amendment proposals.

The Government of Japan has also submitted a draft resolution on the relationship with the International Whaling Commission. Under this draft resolution, if accepted, the Conference of the Parties would have to amend the CITES Appendices I and II with regard to whale species under the criteria of Resolution Conf. 9.24, taking into account scientific information from the IWC and other sources; would affirm that it has not received scientific justification for the moratorium and Southern Ocean Sanctuary, and that, because CITES and the Whaling Convention are independent agreements, any arrangements under CITES would not automatically be linked to the measures taken by the IWC; and would repeal Resolution Conf. 2.9.

As a final remark, which should not be interpreted as a criticism, I would like to say that while whaling nations and/or organizations have for many years stated that the IWC was the organization responsible for whales and that CITES should keep a low profile regarding these species, it appears now that the same nations and organizations are trying to use CITES to their advantage because the IWC has adopted decisions contrary to their interests and does not seem to be ready to change its position. This is an interesting development, which would deserve to be followed!

Whaling and international law

William T. Burke *Professor, School of Law, University of Washington,Seattle, WA 98105   e-mail: burke@u.washington.edu*

The organizers of this meeting have suggested some topics for brief comment and my remarks center on them.

1. A short evaluation of the International Whaling Commission and its present role in view of international law, the Law of the Sea Treaty, and whaling management in general.
2. Actions Iceland must carry out so that its whaling operations accord with the international law of the sea.
3. Can NAMMCO be considered an "appropriate" international organization for the conservation, management and study of whales? If not, what would be required to be "appropriate"?
4. Is it possible for Iceland to reenter the IWC with a reservation to the moratorium?
5. Canada's policy on whaling.
6. Recent IWC decisions and dispute settlement.

1. SHORT EVALUATION OF THE IWC and its present role in view of international law, the Law of the Sea Treaty, and whaling management in general.

The first and overriding consideration in responding to this question is simply what is the international law for whaling. In my view, under general or customary international law the nationals of all states are entitled to harvest whales on the high seas unless the state of registry or flag has agreed otherwise; each coastal state has control over whales within its national jurisdiction, subject to its international agreements, and it has sole authority over the activities of its nationals on the high seas. I think these propositions accurately state current international law on whaling. Some states are parties to agreements that affect their position on whaling both within and outside national jurisdiction. Approximately 190 entities are now generally recognized as independent states, of which 40 have specifically agreed to international regulation of whaling. I am not sure how many of these States have nationals who harvest cetaceans, large and small, but I believe it considerably exceeds 40.

Two major agreements affect whaling for States parties to them, the International Convention fro Regulation of Whaling (ICRW) and UNCLOS, so the more specific question is what is the effect of the ICRW and the UN Convention on the Law of the Sea on the general freedom to take whales. Articles 87 and 116 of UNCLOS affirm that the nationals of all states are free to take living resources on the high seas, subject to their treaty obligations and to UNCLOS itself. UNCLOS articles otherwise do not forbid whaling, nor does the ICRW.

The ICRW seems to me unequivocally to provide for the sustainable harvest of whales. Every article from the Preamble through Article IX envisages the harvest of whales, for commercial, scientific, and cultural purposes, subject to regulation for purposes other than scientific. This view of the treaty is rejected by major parties to the ICRW who believe that it is consistent with the treaty to prohibit all commercial harvesting of whales permanently. There is no basis for such an interpretation that can be considered binding on members who disagree.

Another question is whether principles of customary international law have developed which override the provisions of the ICRW. I am unaware of any such principles. Furthermore, if there were such principles, they would not eliminate invocation of still another principle, namely that persistent objectors to an emerging principle of customary international law are not bound by it. Japan and other whaling members of the IWC clearly object to the view that any alleged customary law principle over-rides the provisions of the ICRW which clearly provide for sustainable whaling.

The most recent international instruments that address the taking of marine mammals, specifically including large cetaceans, do not by any remote feat of the imagination forbid their harvest. As noted above, the

UN Convention on the Law of the Sea affirms the precise opposite of that proposition. A second instrument, although not a treaty, is Agenda 21, the plan of action produced by UNCED. The efforts in Rio de Janeiro to eliminate the right of whaling on the high seas were quite specifically rejected and the right of sustainable harvesting of all marine living resources of the high seas was reaffirmed in para. 17.46. Para. 17.47 reiterates article 65 of the LOS treaty in relation to marine mammals on the high seas, which simply underscores that sustainable harvesting of whales is specifically countenanced unless an individual State or an international organization with the competence to do so limits or prohibits such exploitation.

UNCLOS Articles 65 and 120 do not, in my opinion, require a State accepting UNCLOS (now well over 100 have done so) also to become a party to the ICRW or even necessarily to "work through" that organization, nor do they by themselves modify the right recognized in article 116 to take living resources on the high seas. Article 65 (and 120) uses the plural term "organizations", leaving open resort to other organizations than the IWC when that is deemed an effective way to proceed. I have yet to see a plausible reason for the too common practice of ignoring that plural term.

I am aware of recent suggestions that under current international law whales are now res communis, meaning that they are wholly subject to international controls in the sense that they can only be taken if the international community affirmatively permits it. This view wholly ignores the LOS treaty and its reiteration in article 116 of customary law on freedom of fishing beyond national jurisdiction. One observer finds authority for the res communis nature of whales in the 1970 UN Declaration of Principles about uses of the ocean floor beyond national jurisdiction, arguing that the statement therein of the common heritage of mankind should and can be extended to whales.

I think it can be stated without fear of contradiction that no one in the law of the sea negotiations, which occurred after adoption of this Declaration of Principles, gave any serious thought to including living resources within the notion of common heritage. It hardly takes a genius to recognize that most states in these negotiations were most concerned to capture for their own exclusive use as much of the ocean's living resources as could be coralled in the negotiations. And they managed to get everything in about 35% of the ocean. Marine mammals were not excluded from coastal state sovereign rights. For the area beyond the exclusive economic zone, on the other hand, the notion of freedom of fishing for all living marine resources was accepted, subject to the obligation to conserve.

The IWC itself is obviously not an effective organization in the context of the terms of its original agreement. As noted above, it is my assumption that the relevant provisions (i.e. objectives, scope of authority) of the ICRW have not been changed by the States Parties by means recognized in international law. The ICRW is routinely violated by the majority of its members and no one has yet managed to challenge this, except by withdrawal and sometimes by the objection procedure. But these procedures for registering objections to IWC actions do not change the practice of other States in violating the treaty. In its own terms, in any case, it is a pretty primitive instrument, with no enforcement provisions and no dispute settlement arrangement provided. As is well known, the advice of its Scientific Committee is ignored when it does not satisfy the political preferences of members.

In my opinion, as an outside observer, the IWC could be on the verge of collapse although the conditions for that to happen do not seem likely to occur. Only Japan's unwillingness to withdraw keeps it alive. However, so far as known to me, Japan does not plan to withdraw anytime soon although it mentions the possibility from time to time in reaction to the departures from the treaty.

If Japan did withdraw, and if Norway continues to object to the moratorium, the IWC would become largely irrelevant in relation to whaling. In any substantive sense of regulating whaling, as opposed to trying to halt whaling, it is irrelevant already, but it is propped up by the US threat of embargoes coupled with Japan's unwillingness to force States to abide by the treaty. I have no inside knowledge about Japanese policy, but a guess is that Japan does not rate continuation of whaling very high on the political scale of importance. It will not risk much political injury in defense of its whaling interests.

Current perceptions of the IWC strongly suggest that legal commentary about its actions is mostly superfluous. The IWC is now considered by most of its members as wholly political, resembling the UN General

Assembly in its adoption of resolutions. Legal considerations have little to do with decisions in the IWC. I suspect that this attitude prevails among many members, especially those who must face severe domestic political pressure from environmental groups. There is no countervailing pressure in such states and in the absence of any international pressure it is the easiest decision in the world to indulge whatever environmental groups demand. In my view this describes much of what happens in the IWC.

Thus, in some ways, the IWC resembles the UN itself. The latter has a governing treaty, but it is widely ignored and departures from it are not sanctioned. Even the basic matter of meeting financial obligations is disregarded by the United States, for example, as if the UN Charter itself was wholly unimportant. This is not far from my perception of the US attitude toward the ICRW: its provisions simply don't matter except when it is politically expedient to invoke them.

In relation to "whaling management in general", the IWC really appears to be removing itself from action, given the current refusal to implement the Revised Management Procedure while stonewalling the associated enforcement issues. Simply prohibiting all commercial whaling for an indefinite time is not managing.

Until the moratorium is removed, which some observers believe will not happen soon, the IWC has relatively little to do except play around the edges with discussions of somewhat peripheral issues of aboriginal and small type coastal whaling, the niceties of humane killing, concerns about small cetaceans, and the regulation of whale watching. These matters are not unimportant to those affected as well as others, but to my mind it is somewhat ludicrous to see so much effort and resources devoted to debates about the catch of one or two whales. Ironically, those who complain about these catches are the ones who ignore the treaty in other contexts. So what happens in the end, is that no one can be obligated to do anything unless coerced to do so. This is what results from a pattern of continuous treaty violation. The United States has had a considerable record over the last 15 years of contributing to this state of affairs.

2. ACTIONS ICELAND MUST CARRY OUT so that any whaling will be in accordance with the requirements of the law of the sea.

My understanding of the "requirements of the law of the sea" is that Iceland has the right to engage in whaling on the high seas and the obligation to cooperate with other nations in the conservation of whales. Cooperation might take many forms, of course. Cooperation does not mean that Iceland is obligated to join the IWC by again becoming party to the ICRW. To cooperate with other nations does not mean that it must accept the views of other States that no commercial whaling is permissible until these other States say so. Of course, the ICRW itself has no such requirement.

In one form of cooperation, Iceland can seek to agree, through negotiations with other States, on the conservation measures that are considered necessary for specfic stocks on the basis of the best scientific evidence available. Good faith negotiations would address the needs of whale conservation. Such negotiations would not be dictated by the notion that the only acceptable outcome is an agreement on ceasing all whaling irrespective of stock abundance. This latter is the position taken by Australia, New Zealand, and the United Kingdom in the IWC. Negotiations conducted by States who insist on this outcome to the exclusion of any other goal are not negotiations in good faith. There is no legal requirement of which I am aware that Iceland must negotiate with States who proceed from this premise and postulate this as the only goal for the negotiation.

What national control over whaling is required by international law? There is no obligation under international law to join an international mechanism that deals with whaling although such a mechanism could be helpful if it were employed in good faith (the IWC is not so employed). Article 65 would have to be written in different terms to establish the obligation to join an international mechanism. "Working through" is awfully general language to convey the concept that a State cannot cooperate except by adhering to a specific organization, especially one in which it had no say about its objectives, principles, operations, voting, financing etc.,. If "national control" means national decision-making that seeks to cooperate in the conservation of high seas whales, then I see no difficulty in asserting such control. But, surely, if one can establish an institutional means for cooperation this might make the objective more easy to accomplish. Under current practices and its present organizational structure, the IWC is not "functional" for the purposes of effective conservation action. Iceland, or any other State interested in harvesting whales on a sustainable basis, has every reason to avoid the IWC. Joining it under present circumstances is simply to ask for endless frustration and difficulty. On the other hand, cooperation with the Scientific Committee and use of this information could be a form of productive cooperation.

If the members of the IWC took the treaty seriously and sought actually to regulate for sustainable whaling, then the situation would be very different. The trend of recent events does not suggest this will happen any time soon.

3. CAN NAMMCO BE REGARDED AS AN "APPROPRIATE" INTERNATIONAL ORGANIZATION when it comes to the conservation, management and study of the great whales? If not, what would be required to be "appropriate".

To speak very generally, an "appropriate" organization is one that has the objectives of conservation as the major goal (and this probably needs some more specific definition), has a membership suited to this objective and the interests it serves, has a structure or organization to enable it to work effectively (including research functions, data acquisition, rule-making procedures, surveillance and enforcement provisions, etc.), is provided with the authority to make the required decisions, is funded properly, and has available or has access to scientific information about whale stocks. There are very few, if any, international fishery organizations that meet all of these criteria, but they should be met to the maximum degree possible. (Despite the provisions of the ICRW, the IWC lacks the appropriate objective because in practice a majority of its members does not accept the conservation objective of sustainable harvests of whales) I am not particularly knowledgeable about NAMMCO, but it seems mostly to be aimed at providing a forum for exchanging views and making recommendations that may or may not be accepted.

If NAMMCO is to have credibility at the international level, it probably needs to have its functions spelled out in greater detail, show greater precision in identifying specific regulatory measures and, above all, set up an effective enforcement regime.

4. IS IT POSSIBLE FOR ICELAND TO REENTER THE IWC with a reservation to the moratorium?

The ICRW does not mention reservations and I don't know if others have tried to make reservations. Since the treaty does not prohibit them, at least some reservations could be effective. The moratorium is not itself a part of the ICRW text so a reservation is not specifically needed to combat that directly. While there is little question that a reservation of some kind could be made, its effect would depend upon its subject matter and on the reaction to it of other parties to the ICRW. For those who don't object, an otherwise acceptable reservation would be effective. For those who did object, it would not and conceivably could not allow the reserving state to be a party to the ICRW.

Whatever the legal situation might be, I can't imagine why anyone would want to join the IWC, even with a reservation, since its membership might adopt new and unacceptable measures (how about a sanctuary in the area around Iceland or in the North Atlantic as a whole?) which could only be escaped by an objection, and this invites economic conflict with the US. Given the arbitrary departures from the ICRW by the IWC over the past decade or so, it seems to me remarkable that any State would willingly submit its fate in relation to whale harvesting to such an agency. It must say something about the perception of the IWC that anyone would reach the conclusion just stated. Here is the supposed principal international body concerned with conservation of whales, but its mandate has been so distorted by some members that it cannot be taken seriously any longer as a means of reaching the goal it was established to achieve, namely sustainable whaling.

5. CANADIAN WHALING AND THE LAW OF THE SEA

I have also been asked to comment on Canadian whaling and the law of the sea. My understanding of Canadian policy is based on a 1996 paper by Dan Goodman of the Dept. of Fisheries and Oceans.

"Acknowledging that the International Whaling Commission is responsible for the management of bowhead whaling, it is the position of the Government of Canada that with regard to the harvesting of bowhead whales in Canada, its obligations in respect of article 65 of the UNCLOS to `work through the appropriate international organizations for the conservation, management and study of cetaceans,' are met by working through the Scientific Committee of the IWC."

These statements are not particularly ambiguous, but they may need some interpretation. (1) Canada is not a party to the ICRW, and has no obligation to recognize any authority of the IWC over bowhead whales in Canada or elsewhere; (2) Canada is also not a party to the UNCLOS and therefore has no obligations under article 65 of that agreement to work through the IWC, which in any event is not specifically mentioned in article 65. The thought underlying these statements appears to be that Canada's customary law obligation to cooperate regarding whaling conservation is satisfied by Canada recognizing the role of the IWC in regard to bowhead whales in the Arctic, outside 200 miles at least, and by cooperation through the Scientific Committee. Certainly the seeming acceptance of an obligation regarding either of these agreements cannot rest on the agreements themselves since Canada has not accepted them.

I doubt that Canada's obligation to cooperate requires recognition of the IWC as "responsible for management of bowhead whaling" although such cooperation may be helpful and is certainly appropriate. In fact, it is not clear to me that Canada actually regards IWC as having authority over bowheads since Canada has authorized takes of both western and eastern stocks of bowheads without regard to the IWC. Canada has expressly rejected the resolution adopted by the IWC last June enjoining Canada to obtain IWC approval for permitting bowhead harvests.

6. RECENT IWC DECISIONS, THE ICRW AND DISPUTE SETTLEMENT

I believe that several recent IWC actions are not consistent with the ICRW, including the Southern Ocean Sanctuary decision, the continued arbitrary refusal to implement the Revised Management Procedure, the refusals to allow takes of whales in small type coastal whaling, and the recommendations for undertaking scientific whaling.

I have done some work on the possibilities of dispute settlement under UNCLOS Part XV and the new Tribunal for the Law of the Sea as established in Annex VI of UNCLOS. IWC actions that arguably involve violations of UNCLOS are eligible for compulsory dispute settlement. Disputes about contradictions between IWC actions and the ICRW can also be brought before the Tribunal for the Law of the Sea under Article 288 and Annex VI, but this would require agreement on submission as I understand the pertinent provisions. The significant point here is that there is now an established and recognized body to whom disputes can be submitted. There is no room anymore for the contention that disputes about the actions of the IWC must be decided only by the IWC itself, there being no other alternative.

Recent developments in the IWC aboriginal subsistence whaling category

Ray Gambell *Secretary to the International Whaling Commission, The Red House, 135 Station Road, Histon Cambridge, UK CB4 4NP*

BACKGROUND

The special status of aboriginal subsistence whaling was first formally recognised in a global international treaty in the International Convention for the Regulation of Whaling which was opened for signature in Geneva on 16 January 1931 and came into effect on 16 January 1935. This included as Article 3 the statement that the Convention did not apply to coastal dwelling aborigines, provided that they used canoes, pirogues or other exclusively native craft propelled by oars or sail; they did not carry firearms; and the products were for their own use (Birnie, 1985, pp. 681-2). The first Schedule to the 1946 Convention carried this concept forward by a specific exception to the general ban on the commercial catching of gray and right whales when the meat and products are to be used exclusively for local consumption by the aborigines (IWC, 1950, p.15).

Aboriginal subsistence whaling has therefore been recognised in international treaties for at least 60 years as in some ways different, and having a distinctive character which separates it from the larger-scale commercial whaling operations. The current regulations governing commercial catching activities still include specific exemptions for aboriginal subsistence whaling from the general controls and limitations which they spell out (Gambell, 1993, 101-2).

CRITERIA

Aboriginal subsistence whaling is not formally defined within the 1946 Convention or its associated Schedule of regulations, although provisions within the latter speak of establishing "catch limits for aboriginal whaling to satisfy aboriginal subsistence need" (paragraph 13 (a)).

In 1975 the IWC adopted its new management procedure for commercial whaling and this led to recognition of the need for a separate and specific management regime for aboriginal subsistence whale fisheries (Gambell, 1993, pp. 101-4). This occurred during a period of very contentious discussions on the status of the bowhead whale stock hunted off Alaska and the numbers of whales being harvested by the Eskimos (Doubleday, 1989).

An *ad hoc* Technical Committee Working Group on Development of Management Principles and Guidelines for Subsistence Catches of Whales by Indigenous (Aboriginal) Peoples which met immediately before the 1981 Annual Meeting of the IWC agreed to the following definitions:

Aboriginal subsistence whaling means whaling for purposes of local aboriginal consumption carried out by or on behalf of aboriginal, indigenous or native peoples who share strong community, familial, social and cultural ties related to a continuing traditional dependence on whaling and on the use of whales.

Local aboriginal consumption means the traditional uses of whale products by local aboriginal, indigenous or native communities in meeting their nutritional, subsistence and cultural requirements. The term includes trade in items which are by-products of subsistence catches.

Subsistence catches are catches of whales by aboriginal subsistence whaling operations (Donovan, 1982, p.83).

After consideration of the many biological, nutritional and social aspects involved, the 1982 Annual Meeting of the IWC adopted a Resolution agreeing to implement an aboriginal subsistence whaling regime (IWC, 1983, pp. 28-9, 38).

THE ABORIGINAL SUBSISTENCE WHALING SCHEME

The IWC has recognised that the full participation and co-operation of the affected aboriginal peoples are essential for effective whale management. The catch limits to satisfy aboriginal subsistence need are established in accordance with certain principles. For stocks above the maximum sustainable yield (MSY) level, aboriginal subsistence catches are permitted up to 90% of MSY; for stocks below the MSY level, catch levels are set so as to permit the stocks to rebuild to the MSY level; and the Scientific Committee was asked to advise on both a rate of increase towards the MSY level and a minimum stock level below which whales should not be taken from each stock. A standing Aboriginal Subsistence Whaling Sub-committee of the Commission was established to consider documentation on nutritional, ksubsistence and cultural needs relating to aboriginal subsistence whaling and the uses of whales taken for such purposes, and to provide advice to the Technical Committee for its consideration and determination of appropriate management measures (IWC, 1983, p.28-9, 38, 40).

Now that the scientific component of the RMP for commercial whaling has been completed, the IWC has asked its Scientific Committee to begin a review of possible alternative management regimes for aboriginal subsistence whaling (IWC, 1995, pp.22, 42-3).

PRESENT ABORIGINAL SUBSISTENCE WHALING

There are four whale hunting operations presently regulated by the IWC as aboriginal subsistence whaling activities and for which it sets catch limits:

Bowhead whaling off Alaska. The present aboriginal subsistence whaling regime is implemented for the Alaskan bowhead hunt by the Government of the USA calculating subsistence need as the historic harvest per capita of the human population involved, multiplied by the current human population. The IWC has taken this figure as the basis for setting the catch limit as 204 bowheads to be landed in the four years 1995-1998, reflecting the estimated annual Eskimo population growth rate for the ten whaling villages concerned of 4.7% between 1990 and 1992. This catch limit is less than the estimated current annual replacement yield from the stock of 199 animals (95% confidence intervals 97 - 300) (IWC, 1995, p.21-2).

Greenland whaling. Humpback whales were a traditional quarry for whalers in Greenland. The general protection of humpback whales from 1955 carried an exception for Greenlanders using small vessels until 1985. The local minke whale fishery began in 1948 in West Greenland. There is an annual catch limit of 12 minke whales for the people of East Greenland. The present quota of 19 fin whales landed and 165 minke whales struck in each of the years 1995-1997 in West Greenland is equivalent to a yield of 420 tonnes of meat and products but this is less than the accepted need for 670 tonnes by the local population in West Greenland (IWC, 1995, p.22).

Siberian gray whaling. The present limit is 140 gray whales for each of the years 1995-1997 which corresponds to the catch requested by the Government of the Russian Federation (IWC, 1995, p.22). This can be compared with the estimated annual replacement yield of 611 whales with 95% confidence intervals 452 - 786 (Allison et al, 1995).

Caribbean humpback whaling. A small-scale open boat hunt carried out from the island of Bequia in St Vincent and The Grenadines in the Caribbean continues with an annual quota of two humpback whales, which are not always taken (IWC, 1997 in press).

INTERPRETATION

The variety of attitudes which exist in the IWC when it comes to considering catch limits for aboriginal subsistence whaling, in addition to the hunts described above, can be illustrated by the discussions which took place at the 49th (1996) Annual Meeting of the Commission.

Chukchi bowheads
At this meeting the Russian Federation presented a request for an annual catch of 5 bowhead whales to meet the needs of the indigenous people of the Chukotski Autonomous region. The USA supported this request as fully justified under IWC cultural and subsistence needs criteria for aboriginal whaling.

However, a number of delegations sought clarification about the needs of the Chukotka people, given the under-utilisation of the existing quota of gray whales, while others urged caution, because of the endangered state of the bowhead stocks. The Russian Federation responded that economic changes experienced throughout the Russian Federation had impacted on the region and on its food security, temporarily disrupting whaling operations which were now carried out by the whaling villages themselves. The new quota would supplement gray whale meat and was also required for ceremonial and cultural purposes. The current bowhead whale catch limits reflect the needs of other populations in other countries and it believed that it was not appropriate for the IWC to meet Chukotka need in ways detrimental to others. The Russian Federation confirmed that bowhead meat would not be used in fox farms and was solely for human consumption, and while the requested quota would not fully make up the deficit in gray whale meat, it would improve food security.

The Russian Federation took the view that decisions on aboriginal subsistence whaling should be taken by consensus, not by vote as is the case for commercial whaling. Because there was no consensus it indicated that no vote was needed and there would be no request for the item to be considered further (IWC, 1997).

Makah gray whales
The USA presented a request for a catch of 5 gray whales by the Makah Tribe which lives on the Pacific coast in Washington state. Although the commercial exploitation of gray whales in the late 1800s had led to the suspension of whaling by the tribe since 1926, it noted the continuance of aspects of the whaling tradition within the tribe since that time, and it emphasised the strong community and tribal aspects of the whaling proposal.

While some delegations were fully supportive of this proposal, others expressed reservations because of the 70 years of non-whaling which suggested that there was no clearly demonstrated need. Concern was also expressed over a widening of the scope of whaling activities at a time when whaling was thought to be coming to an end, the commercial element in the Makah whaling, and apparent divisions within the tribe itself.

After consultations with Makah representatives the USA withdrew its proposal and asked the Commission to defer consideration until next year when the gray whale quota expires and the needs of the Chukchi people will also be determined (IWC, 1997).

Canadian whaling
Another factor in the management of aboriginal subsistence whaling highlighted by the USA at the IWC's 1996 meeting was the whaling activities of non-member states. In particular, the possibility of a hunt for bowhead whales in Canada was mentioned. This gives rise for concern both because of the small size of the Davis Strait and Hudson Bay stocks, and the issue of whaling outside IWC control generally. The Canadian observer noted that the aboriginal people of Canada have a constitutionally protected right to harvest fish and marine mammals subject to conservation, and recalled that his government had banned commercial whaling in 1972 and withdrew from the IWC in 1982 concluding it had no reason to remain as a member of the IWC which is mandated to make possible the orderly development of the commercial whaling industry. The aboriginal harvest in no way represents a re-initiation of commercial whaling activity. Canada felt strongly that a Resolution adopted by the IWC encouraging it to reconsider the issuing of permits for such catches and to re-join the IWC if it continues to have a direct interest in whaling, was negative, inappropriate and counter-productive (IWC, 1997).

It may be noted that there is at least a theoretical possibility of a single stock of whales being harvested commercially and under the IWC's aboriginal subsistence whaling scheme, and by a non-member government of the IWC. Such a situation will need careful consideration and management.

Matching with criteria

The reservations apparent in the IWC over any additional aboriginal subsistence whaling can be related to the unwillingness of the majority of governments to permit a resumption of small-type coastal whaling by Japan and Norway.

Since the 1982 ban on commercial catching of whales came into effect the Government of Japan in particular has made long and strenuous efforts to gain recognition of, and to alleviate, the distress caused in its coastal communities which had formerly relied heavily on small-type whaling. It has presented much documentation on social, scientific and anthropological research supporting the conclusion that Japanese small-type whaling has a character distinct from other forms of industrial whaling and sharing some of the features of aboriginal subsistence whaling. The small-type whaling in Japan is a small-scale limited access fishery involving four coastal communities, taking minke whales within 30 miles of the shore. The whale meat obtained from these catches is claimed to play an important role in the cultural and social cohesion of the communities (Institute of Cetacean Research, 1996).

Norway has also presented evidence that although modern minke whaling was introduced in the 1920s, small cetaceans have been hunted in Norway for millennia. The whalers are also fishermen, generally based on household units of ownership and crew, which gives strong support to the traditions and way of life of the remote northern communities. There is considerable resentment against the prohibition of catching activities imposed by other members of the IWC (IWC, 1993, p.16).

Japan and Norway have argued that the ban on the small-type whaling operations from their coastal communities, because they are at present classified by the IWC as commercial whaling, is unjustified. It is causing problems in the communities because of their dependence on the social and cultural activities associated with the whaling operations and the distribution and consumption of the whale products. These aspects are shared to a significant degree by the aboriginal subsistence hunts which are not prohibited by the IWC because of its recognition of their special socio-economic and cultural roles in the lives of the northern communities concerned.

The difficulty for some other governments in accepting these claims lies in the commercial aspects of the small-type whaling operations, even though the catches taken under the aboriginal subsistence whaling arrangements in both Greenland and St Vincent and The Grenadines undoubtedly are sold commercially, and the Alaskan and Siberian operations must also involve money transactions to cover the costs of equipment and supplies.

Requests for an interim relief allocation of 50 minke whales for the

Japanese operations, regulated and controlled through application of the Revised Management Procedure and involving an Action Plan approved by the IWC for a strictly non-commercial distribution system for the products, have not been accepted by the IWC (IWC, 1997).

Following its failure to persuade the Commission to de-classify the Northeastern Atlantic stock of minke whales (IWC, 1992, pp.26-7) to allow the resumption of the traditional Norwegian coastal whaling where there is a demonstrated cultural and subsistence need, Norway has set its own quotas since 1993 for limited catches. It is able to do this legally because it lodged formal objections to both the Commission's ban on commercial whaling and the classification of the Northeast Atlantic stock of minke whales as a Protection Stock under the NMP at the time these measures were adopted. A total of 581 minke whales has been killed in the last three seasons.

REVISION OF THE MANAGEMENT SCHEME FOR ABORIGINAL SUBSISTENCE WHALING

The Scientific Committee is continuing the work started in 1995 to draw up an Aboriginal Subsistence Whaling Procedure. The objectives and rationale of the procedure can be summarised as:

(1) to ensure that the risk of extinction to an individual stock is not seriously increased by subsistence;

(2) to enable harvesting in perpetuity at appropriate levels;

(3) to maintain the stocks at or above an optimum level (giving highest net recruitment), or if they are below it, ensuring that they move towards that level;

(4) highest priority will be given to objective 1.

The Scientific Committee is also addressing the issue of need, and has sought guidance from the Commission on what is expected from the Scientific Committee. It seems desirable to develop a model to demonstrate the possible effect of various levels of need, and the Commission has emphasised the importance of feedback from the native hunters and the Commission.

Where possible the Scientific Committee is using the same performance statistics as were used in the trials for the RMP, particularly with respect to risk. However, the differences in the objectives for an aboriginal subsistence procedure and the RMP mean that it is not possible to use an identical set of statistics. As part of the process the Scientific Committee is considering the current aboriginal whaling scheme and variants of the RMP, but is not limiting itself to these options (IWC, 1997).

CONCLUSION

There is a general view within the IWC that aboriginal subsistence whaling has to be seen within the broad concept of sustainable use. A number of governments, including Denmark and the USA, have stated explicitly that the current system is working quite well (IWC, 1997). There is therefore a degree of hesitation in making changes unless these can be seen as clearly representing improvements over the present arrangements. A particular factor thought to be especially important by a number of governments is the evidence of cultural continuity in the hunt and its role in the societies involved. Potential conflicts also arise when the traditional methods of killing may not be as humane as could be achieved by the introduction of improved modern technology. However, the experience of the introduction of a new grenade containing penthrite explosive, and associated aids to tracking and locating bowhead whales in the Alaskan hunt, illustrate the possibility of retaining the traditional hunting methods and culture and at the same time improving humaneness and efficiency (Øen, 1995).

ABSTRACT

Aboriginal subsistence whaling has been recognised in international treaties since 1931, and although not defined is recognised in the 1946 International Convention for the Regulation of Whaling. The IWC developed a management regime for aboriginal subsistence whaling in 1982 which is distinct from the management procedure designed for commercial whaling. This includes recognition of the nutritional, subsistence and cultural needs of the aboriginal peoples affected. Bowhead whaling off Alaska, whaling from Greenland, Siberian gray whaling and humpback whaling in the Caribbean are regulated by the IWC.

A bowhead catch by the indigenous people in Chukotka and a take of gray whales by the Makah tribe were requested at the 1996 Annual Meeting of the IWC. These were supported by most member governments but others expressed reservations because of worries including the increasing catches and lack of continuity in the catches, and both proposals were withdrawn. Catches by Canadian aboriginal people, outside IWC control, also causes concern. In addition, small-type coastal whalers in Japan and Norway who are subject to the ban on commercial whaling argue that they have cultural traditions similar to the aboriginal people who are permitted to whale.

The aboriginal subsistence whaling management scheme is now being revised, although some governments think the present scheme is working well and there is reluctance to make changes. Cultural continuity in the hunts is thought important, as well as improvements in the humaneness of the killing methods.

REFERENCES

ALLISON, C., Punt, A. E. & Butterworth, D. S. 1995. Census Based Estimates of RY for Gray Whales. *Rep. Int. Whal. Commn.* 45:162.

BIRNIE, p. 1985. International Regulation of Whaling. *Oceana Pubs.*, New York.

DONOVAN, G. P. 1982. The International Whaling Commission and Aboriginal/Subsistence Whaling: April 1979 to July 1981. *Rep. Int. Whal. Commn* (Special issue 4);79-86.

DOUBLEDAY, N. C. 1989. Aboriginal Subsistence Whaling; The Right of Inuit to Hunt Whales and Implications for International Environmental Law. Denver J. *Int Law and Policy.* 17(2):373-93.

GAMBELL, R. 1993. International Management of Whales and Whaling: An Historical Review of the Regulation of Commercial and Aboriginal Subsistence Whaling. *Arctic* 46(2):97-107.

Institute of Cetacean Research, 1996. Papers on Japanese Small-Type Coastal Whaling Submitted by the Government of Japan to the International Whaling Commission 1986-1995. Japan. Pp.296.

International Whaling Commission. 1950. International Convention for the Regulation of Whaling. *Rep. Int. Whal. Commn.* 1:9-14

International Whaling Commission. 1983. Chairman's Report of the Thirty-Fourth Annual Meeting. *Rep. Int. Whal. Commn.* 33:20-42.

International Whaling Commission. 1992. Chairman's Report of the Forty-Third Meeting. *Rep. Int. Whal. Commn.* 42:9-50.

International Whaling Commission. 1993. Chairman's Report of the Forty-Fourth Annual Meeting. Rep. Int. Whal. Commn. 43:11-53.

International Whaling Commission. 1995. Chairman's Report of the Forty-Sixth Annual Meeting. *Rep. Int. Whal. Commn.* 45:15-52.

International Whaling Commission. 1997. Chairman's Report of the Forty-Eighth Annual Meeting. Rep. *Int. Whal. Commn.* 47: in press.

ØEN, E. O., 1995. A New Penthrite Grenade Compared to the Traditional Black Powder Grenade: Effectiveness in the Alaskan Eskimo's Hunt for Bowhead Whales. *Arctic* 48(2):177-185.

Fostering a negotiated outcome in the IWC

Robert Friedheim *Professor of International Relations, School of International Relations, University of Southern California, VKC 330 Los Angeles, California 90089-0043 e-mail: friedhei@rch.usc.edu*

THE NATURE OF THE PROBLEM

If one wished to succeed in bargaining in the International Whaling Commission (IWC) one would have to bargain smarter, not harder. But the strategy and tactics used by both pro- and anti-whaling states, state coalitions, and nongovernmental organizations has produced stalemate, and unless some of the major players break from their long-held positions, stalemate will be the best that can be hoped for in the foreseeable future.

This is not to say that the major actors are happy with the present stalemate, but they fear they will be worse off if they shift their position or take another approach to the problem. After all, the major players have *some* of what they want under the present arrangement.

The issue is not ripe for resolution by negotiation. As noted, the majority hopes to impose an essentially legislative outcome on the minority. If they continue on this path, they will succeed either by outlasting the minority, or by coercing it. But thus far the minority has been obdurate — they will not concede. While the stalemate may last for some time into the future, it is unlikely to last forever. But someday the specter of defection or leaving the negotiation will loom large, because some of the participating states will believe that they will be better off with no agreement than an unacceptable agreement.

Ripeness as a concept: Its Application to IWC Bargaining

"Ripeness" is a negotiating state that many experienced diplomats recognize when it happens, but few are in a position to force it. It is defined as a "moment of seriousness," a turning point where all stakeholders recognize that it is actually possible to arrive at a solution to the problem by a joint decision. Another way of putting it is that the parties realize they will be better off by agreement than disagreement and therefore cooperate in shaping a joint outcome. But it often comes about because of shift in factors over which the negotiators have little control. Therefore it is quite difficult to predict when it will happen, or what might be done to cause it to happen. In the IWC, negotiators face a situation of "positional" bargaining, where the stakeholders insist on the inviolability of the principles which undergird their positions, are are therefore unwilling to look at a proposed solution unless it fits squarely within their conceptual framework. This creates stasis.

But stasis does not last forever. Historical observation shows shifts on many issues under negotiation that result from changes in exogenous factors. More often than not, stasis collapses of its own weight. The "real world" problem under negotiation has changed while the proposed solutions have remained fixed. At some point the parties recognize that what they are doing by sticking to their position is irrelevant. A new solution is needed.

Sometimes ripeness occurs out of sheer exhaustion. All sides are tired and want to end the constant contentiousness. They become more "reasonable." But I think that reasonableness can be fostered if the parties, reaching a state of fatigue, have available a set of ideas that look at the problem in the light of new circumstances. In short, they see a way out of the present dilemma.

In the meantime, despite pessimism that more appropriate tactics alone can turn the situation around, I offer below some observations on the nature of the bargaining problem and some recommendations concerning tactics useful to the circumstances. Strategy, tactics, and leadership count. It might be possible to get a better outcome by bargaining smarter. I will borrow freely from a large literature and show how the ideas I adapt can illuminate IWC discourse. We will begin by looking at the IWC as a bargaining arena that creates both constraints on the bargainors and opportunities for creative resolutions.

THE IWC AS A BARGAINING ARENA

The International Whaling Commission, meeting annually, and making decisions binding upon most of its members, is an arena for large-scale, multilateral negotiations. It is arguably a real negotiating venue, since decisions can be taken only with the consent of the affected parties. If they do not consent, a party feeling itself to be made worse off by a joint decision can leave the organization, enter an objection to a decision to which it refuses to be bound, or partially avoid being bound by using a right (e.g., to conduct scientific whaling and continue to use the whales caught for human consumption) to avoid an obligation, such as a moratorium on commercial whaling. Negotiations occur at many levels within the IWC. These negotiations range from cooperative to very adversarial.

The IWC shares with other large-scale multilateral negotiations a number of common characteristics. But it also has a number of distinctive attributes of its own which very importantly affect the outcomes of its proceedings. Most of the major common attributes of multilateral negotiation have been summarized by I. William Zartman: (1) multiple parties, (2) multiple issues, (3) multiple roles played by participants, (4) its purpose is rule-making, and (5) decision process dominated by coalition formation. However, the IWC is an "outlier" to other major multilateral negotiations because its decision process is not consensus-driven (defined as a coalescing around a preferred outcome of a dominant coalition with dissenters — if any —acquiescing by going silent; thus unanimity is not required). In addition, even where it shares general characteristics with other multilateral negotiations, it has shaped some of these characteristics in a way quite different from other negotiating venues.

All the attributes of multilateral negotiations add up to a whole noted most for its complexity. Many negotiators, many stakeholders (those who have a "stake" in the outcome), many issues, often technically complex, and the difficulty of integrating the preferences of multiple actors into a commonly accepted outcome, continuous negotiations over a prolonged period of time, and preparatory work performed by an international organization. These force the use of simplifying mechanisms for gaining outcomes such as coalition. They also can limit the scope of outcomes, since often it is the "least common denominator" outcome that can gain support under a consensus requirement.

The IWC has all of these attributes. State participants are quite asymmetric in terms of the usual attributes of political power, including the remaining hegemon, the United States, as well as Japan, the major European states such as the Russia, United Kingdom, France, Germany, a number of middle powers, and a number of small countries. As a result, important roles are often delimited by the power of the participating state. Nevertheless, since all have a right to participate, be heard, and vote, and all have to be accounted for in the decision process, decision-making is cumbersome. Even more telling in recent years, the nongovernmental stakeholders also have to be heard, and indeed, have a right to participate in virtually all phases of IWC's business.

The IWC also seemingly deals with many issues, or at least deals with one issue with many manifestations. There are numerous working groups, workshops, and committees dealing with a revised management program and scheme, catch limits, moratoria, sanctuaries, humane killing, management of small cetaceans, aboriginal subsistence whaling, whale watching, and data collection and analysis. Sessions of the IWC are busy times. But because all IWC's issues are highly interrelated, it is difficult to promote inter-issue tradeoffs, and in this regard, the IWC is different from many other multilateral negotiations that try to resolve a wider variety of issues.

The IWC deals with issues that require specialized knowledge. Therefore persons with specialized knowledge have always played an important role in IWC decision-making. But lately there is tension between key groups. In an earlier period it was whaling industry managers who controlled the data on whale catches and industry profitability that often moved decisions. More recently, it has been scientists who have not only provided technical expertise concerning the nature of the problem, but also provided recommendations concerning its solution. Scientists expect that their recommendations will be taken seriously and translated into authoritative decisions. When they are not, some important scientists become frustrated.

Recent social science scholarship on international environmental decision-making has pointed out, in the cases studied, that when scientists form a consensus on the cause and solution of a problem, they participate in an "epistemic community" and their unified view drives the outcome. However, other cases, including the present one, limit the "generalizability" of this observation. Based on the whaling case, Ronald Mitchell, suggested that a scientific consensus demonstrating an activity's instrumental harm strengthens support for environmental measures, while a scientific consensus demonstrating the absence of harm will weaken support for environmental protection. Organized environmental groups and governments committed to a "moral" solution pay attention to concerns about harm and ignore observations showing no harm, as they did on the question of establishing a Southern Ocean Sanctuary.

As in many other multilateral negotiations, the interactions of its participants is on a periodic basis. Since it is not a one-time meeting of the parties there is a "shadow of the future." Because they may meet again, although the future is discounted relative to the present, no governmental negotiator wants a reputation based on present tactics which might impair his/her ability to function effectively in the future. Therefore deceptive tactics, bold threats, personal attacks, etc., are rarely used by governmental representatives. This observation does not apply to representatives of some NGOs who do not seem to worry about the shadow of the future.

There is a good deal of continuity of people and issues year-to-year. Thus the IWC has some internal stability, and there is substantial momentum to lines of policy development put in motion. But the flip side is also present — there is little urgency to resolve a problem in the short run since everyone knows it can be considered next year. The internal atmosphere of the IWC is a curious blend of the stable and unstable, the orderly movement of business through the agenda, and a three-ring circus.

The meetings are well served by a small Secretariat that is technically competent and adroit at being able to get along with the contending parties. However, the overall organization is quite fragile from a financial point of view. It depends upon the assessments of a limited number of states, some of whom claim to be damaged by the decisions of the majority. If the minority were to withdraw their financial support, the IWC would be badly damaged. There is no way of knowing whether the majority would make up a short fall, but some observers doubt that they would in the light of major developed states' demands to make international organizations leaner.

The decisions sought are rule-making decisions concerning access, allocation, use of a scarce natural resource, or as others see it, a unique world treasure. But in practice, the core of the IWC's business is amending a "Schedule" in which whale stocks are classified, making rules concerning whether a stock can be exploited and to what extent (a quota), and if it can be captured by what capture rules, and how the entire process can be supervised and controlled. In addition, an increasingly important part of its annual business is passing resolutions concerning all aspects of whales and whaling in order to provide guidance (if one wishes to be polite) or to put pressure on (not polite) the parties to conform their behavior to the resolution's demands. These resolutions have no formal standing in international law, and in theory are not binding upon the parties. But those who endorse many of the resolutions hope they will become part of "soft law" and, since they claim they reflect commonly accepted norms, will be obeyed.

The principal difference between the IWC and many other international organizations is the fact that the IWC's dominant decision system is parliamentary while the others rely upon consensus rather than votes for decision. Decisions in the IWC are by majority positive vote. In the strictest sense, one might characterize the IWC as legislative rather than a negotiating system.

The underlying problem is state sovereignty. Even if the world political system is considered less than anarchic, political entities called nation-states still believe they have an attribute called sovereignty so that on important issues they are bound only by those measures to which they consent. Consensus-based international organizations or conferences are a concession to that perception of sovereignty. The IWC's founding fathers in 1946 also made their bow to sovereignty, but in a different way. They created Article V(3), which allowed a member state to file an "objection" to a Schedule amendment so that it can opt out from a rule accepted by a majority. Article VIII also gives member states an absolute right to conduct scientific whaling under special permits and to process the whales caught. Further, Article VIII states that "the killing, taking, and treating of whales in accordance with the provisions of this Article shall be exempt from the operation of this Convention."

Representatives of a majority in the IWC have either overlooked the underlying sovereignty problems, or have a "hidden agenda" and are using the IWC as a venue for attacking sovereignty. For some stakeholders, saving whales is the objective, for others reducing sovereignty is what they are after, and for still some others, these ideas reinforce each other and form what has been called an "overlapping cleavage." As a result of this buttressing, no compromise is possible. IWC politic has evolved toward the parliamentary with states behaving like political parties in a domestic legislature, but not having the underlying social compact which assures that the minority will bow to majority policy demands. Behavior is crudely majoritarian. An anti-whaling majority has controlled the general policy direction of the whaling regime, but has not been able to control the behavior of some important dissenters. It is clear that an anti-whaling majority sets the agenda, does not allow its opponents any victories, and indeed, harasses them through resolutions to give up their sovereign rights. Since they have not acquiesced completely in the face of this assault, it can be said that the present whaling regime is not a negotiated regime.

The present whaling regime is a coercive regime depending upon enforcement measures not agreed to by the minority. The chief public enforcer is the United States through the Pelly Amendment to the U.S. Fisherman's Protective Act. It authorizes the President to certify any state as being in noncompliance with the Act whose actions diminish the effectiveness of international environmental agreements. Miscreants can, in the first place, have their fishing rights in the U.S. Exclusive Economic Zone cut or eliminated, and if that is not enough, they can be subject to unilateral trade sanctions. The former threat has been hollow for some years, since foreign fishing quotas in the U.S. 200-mile EEZ have been eliminated for other policy reasons, but any real imposition of trade sanctions may well place U.S. sanctions squarely in the sights of the World Trade Organization dispute-resolution procedures. But thus far the Pelly Amendment has served as deterrent even though actual punishments meted out have been mostly slaps on the wrist. On the other hand, private coercion through consumer boycotts perhaps have had more impact.

A critical feature of all multilateral negotiation is its dependence upon the formation of winning coalitions to mold outcomes. This is especially true of the IWC, where the process more nearly resembles legislative than bargaining behavior. But in any organization in which three or more parties are trying to devise an outcome acceptable to all, the first task is to persuade one other party that you should create a common position because it advances your common interests. Then you can go into the next phase. If the second phase is legislative, you outvote the third party. If it is a bargaining phase, you try to create a consensus based upon the formula you already worked out to persuade the third party to join the other two.

Coalition opens up a rich lode of tactical maneuvers. The purpose of all of these is to build and defend a winning coalition that unites around a particular substantive formula. To do so, since about the middle of the 1970s, key members of a group of states opposed to all forms of whaling, used a bandwagon effect, logrolling, and other such measures. It is now a dominant coalition and has changed little in composition since it was formed. It has held together in all the various sub-issues relating to whales and whaling — unlike coalitions in many other multilateral negotiations.

A coalition must have a "formula" idea to back; a principle of justice, a set of referents or underlying values that give meaning to the items under discussion. This is discovered in a diagnostic phase, it moves through establishing the formula phase, and culminates when all details are worked out in a final phase. Ending whaling is the formula idea that has appealed to the broadest number of members of the IWC. It clearly purports to be a principle of justice, and it was established as the formula idea of the majority in a second phase before the 1982 moratorium vote, and it has been subsequently refined in a third phase that culminated in the Southern Ocean Sanctuary vote.

The coalition backing the no-whaling formula has shown no willingness to accept anything but its maximum demands. If a negotiating party disagrees with the dominant coalition's formula, it has only two choices if it chooses to remain in the negotiating arena and not defect: (1) try to break up the dominant coalition; that is, replace it with its own new, dominant coalition espousing a different formula idea, or (2) try to get the dominant coalition to mitigate its demands, that is, accept less than a maximum version of its policy idea. Both are difficult and usually costly strategies, but the attempt to form a new dominant coalition is the more costly. But it might be possible. The other alternative is less costly, but it remains to be seen if it is any more feasible than smashing the dominant coalition. Mitigation efforts begin with naysayers recognizing that the dominant formula notion will represent the main policy thrust in the issue area. But if the majority is willing to mitigate, minority participants willing to accept the formula might ask the majority for vaguer language under which the minority can still maneuver, exceptions to the rule, application of the formula notion only to specific geographic regions, extensions of time for application, different language which makes it at least appear that they have not surrendered, or some form of compensation for acquiescing.

These are much easier in a multilateral negotiation with a variety of issues. As noted, IWC negotiations are characterized by a limited range of issues — all relating to whales and whaling. In many other multilateral negotiations the major effort goes toward constructing a grand package which includes tradeoffs within it. There is no grand package in the IWC. As far as I can tell, there are no major tradeoffs. There is an expectation of "sincere" rather than strategic voting. A sincere vote is one where the voter makes his/her choice on the merits of the issue regardless of how it might affect other interests. Strategic voting is when the voter votes not on the merits of the issue but on the basis of whether the measure's supporters in turn support or oppose the strategic voter on his/her salient issues. Despite the lack of variety in issues, issue linking may be possible.

Not unique but quite distinctive to the IWC is the degree of participation of NGOs in the decision process. One side perceives that the issues are largely moral and the other side that the preferences of the majority would unfairly destroy cultures and livelihoods. This accounts for much of the bitterness of participants on all sides. Whaling is an extremely divisive issue. It has resulted in a situation of close to zero trust. As a result, at this stage, it is virtually impossible to counsel that the parties try to move the discussions from the strictly distributive (wherein dividing a "pie" one side's gains add up to the other side's losses) to the integrative (where both sides cooperate to increase the pie before it is divided). Each sees the other as immoral and untrustworthy.

The anti-whalers are viewed by the pro-whalers as racists and madmen. They are seen as trying to impose as a universal standard a set of "ethical" requirements based on western urban values. Anti-whaling spokespersons are not seen as "responsible"; they are viewed as fringe members of western urbanized societies. The rhetoric of the pro-whalers is filled with the hope that some day important developed countries will "wake up" and sweep aside the lunatic fringe members of their societies who have captured that county's position on whaling.

On the opposite side, the anti-whalers combine several major viewpoints which allow all the subgroups of the anti-whaler coalition to come together to attempt to ban all whaling forever. Many non-normative radical observers of whaling view the prowhalers as untrustworthy predators with no sense of limit. They cannot be trusted not to overreach because they have a "bad" history. They should not be allowed to begin the cycle again by permitting commercial whaling to resume. They may be amenable to nonadversarial discussions of the issues, but their suspicions make genuine negotiations that might lead to a bargained outcome very difficult. But their coalition partners are much more difficult to bring to the bargaining table. They are true believers who might be characterized as a religious left who insists that there is a new moral standard, applicable to all. Compromise is not the way to bring the dissenters around. They must be coerced into conceding that whales are sentient beings with rights. As with all true believers, tolerance for the position of others is in short supply.

I do not mean to imply that the pro-whalers have been especially tolerant, or that whalers in the past did not get out of control, but the situation today is a product of history. That history changed fundamentally after the structure of the bargaining arena changed. In 1978, the United States proposed that all IWC sessions and not just plenary sessions be opened to participation by NGOs. With a right to participate assured, it was worthwhile for numerous NGOs to join. More states also joined and it has been said that some were provided inducements to join, or once members, were provided further inducements to assure their votes. It is difficult to negotiate in a bargaining arena with over 90 NGOs intervening in all parts of the process; with over 35 states that have to be brought to consensus if there is to be a negotiated outcome.

The above description should demonstrate what the IWC is *not* — it is not a perfect democratic meeting place. There is very little effort to persuade, only to outvote. There is very little effort to listen, only to repeat arguments.

ADVICE TO THE PRINCES

What can be done to bargain smarter? Can outsiders come up with a fresh approach? I think so, even if some of the points raised dismay some present negotiators. There are few new ideas that will not raise the price of solution for some stakeholders — states and NGOs. Indeed, an "advisor" would have to be Machiavellian in approach, prescriptive rather than descriptive; the bold advisor would have to offer advice to the "Princes." Note the plural. I believe both sides would benefit by rethinking the problem if there is to be a negotiated solution, although obviously the losing side would gain more from movement away from the present situation than the winning side.

Twelve points will be made below to provide some ideas about how to break the stalemate. These categories are for the convenience of presentation. Many of the points overlap, so theoretical neatness is lacking, but they may provoke some of the stakeholders to rethink a situation described by David G. Victor and Julian E. Salt as "empty law."

Salience

There will be no movement until the salience of the whaling issue changes among the stakeholders. By salience, I mean a preference for achieving a favorable outcome on one issue as compared to a favorable outcome on other issues valued by a stakeholder. Salience is always a comparative notion; it is a measure of relative importance. Almost all individuals and collectivities have multiple objectives in the world they face. Except in the rare instances where they value each of their objectives equally, they usually favor one or some over others. When stakeholders discover they cannot win favorable outcomes on all issues, they are forced to accept a tradeoff — a less favorable outcome on an issue of lesser salience in return for a more favorable outcome on the issue of higher salience. While there is some hope of persuading another that your preferred outcome is substantively superior there is a better chance of changing their perception of how much it will cost to achieve a particular objective. Salience can be manipulated. But it may be even more important —albeit difficult — to reexamine how much one is willing to pay to achieve one's own objectives. In effect, one must demand that other internal stakeholders accept a tradeoff of some of what they value so that you have a better chance of gaining what you value. In many cases, domestic discussions and negotiations to change national salience must be concluded successfully before one can negotiate externally to change others' salience, but such internal negotiations can be especially painful.

As it stands, the stalemate on whaling reflects both normative notions of the correctness of allowing or preventing whaling and notions of how much it would cost the antagonists to prevent or restore whaling. Until these perceptions are changed, the protagonists' policies and their bargaining stances will not change. Norway and Japan have said that whaling is an important issue but have not demonstrated a willingness to pay a higher price than they presently pay to achieve it. Conversely, major antiwhaling states claim that ending whaling is a matter of great substance to themselves. But they pay a low price to achieve it. One may speculate on how much they would be willing to pay if others raised the costs to them. They have no whaling industries, few internal pressure groups that could influence them to change. They have to consider only external costs, and these have proved to be low. They can avoid being forced into a legal solution by denying jurisdiction. With the exception of New Zealand, the only way to change their salience is externally .

Pro-whaling states must increase the risks inherent in tougher policies. They must create a sense of urgency for compliance with their demand. If done properly, there should be positive as well as negative inducements (to be discussed in several other categories). One need not threaten war, a breaking off of relations, etc., only that the other party will have to pay a higher price to achieve their objective at your expense. There are multiple objectives even in the whaling negotiations. The anti-whaling coalition must be reminded that the price of obduracy may be dissolution of the IWC. The destruction of the IWC would be a very bad precedent for other international environmental negotiations, a demonstration of where "empty law" could lead. Unless the present winning coalition sees that some losses are possible, they have no reason to give up their gains.

Leadership

The IWC sorely lacks leadership. I do not mean that competent persons have not led their delegations or NGO participants in efforts to fulfill their mandates, or competent international civil servants from the IWC or other participating intergovernmental organizations have not done their duty well. They have; they have pushed for, and some have achieved, the outcomes consistent with their instructions — but at the expense of finding a negotiated solution. One or more persons must emerge who attempt to bring the parties together to accept a solution that makes them all better off by agreement than by disagreement. To do this, they must go beyond the narrow mandate of the stakeholder organization that provides them official credentials.

Several types of leadership in institutional bargaining have been identified. Oran Young and Gail Oshrenko note three: (1) structural; (2) entrepreneurial; and (3) intellectual leaders. Arild Underdahl adds three others: (1) unilateral; (2) coercive; and (3) instrumental leaders. While the overlap is not perfect, it can be argued that there is a substantial similarity between structural and unilateral leaders and between entrepreneurial and instrumental leaders.

I think in recent years in the IWC, structural leadership by the major developed states of North America, Europe, and middle states of Oceania have been exercised to use their material advantage to crush the pro-whalers, who, with the exception of Japan, are smaller or materially weaker states. This has been combined with important intellectual leadership provided by, for example, Dr. Sidney Holt and the spokespersons of such organizations as Greenpeace, who claim they have moved world opinion to the side of no whaling by the quality of their arguments.

The ideas of the anti-whalers are not the only ideas. If the pro-whaling hope to have any chance of working out a negotiated solution, they must provide stronger intellectual leadership. They must have ideas — ideas about both the substance and process of solution. They must produce individuals who can create a sense of trust in their ideas. Without that sense of trust, ideas are just ideas. But more is necessary. If there is an attempt to lead, some coercive leaders will have to emerge. By coercive leadership, Underdahl does not mean threats to mobilize one's army over an issue, but threats and rewards, carrots and sticks. It seems to me that relatively few carrots and sticks have been displayed in the IWC.

I hope entrepreneurial (or instrumental) leaders will emerge. But I cannot suggest how how to bring one or more of them forth. Clearly, the IWC would benefit from the emergence of someone trusted by all sides, who had convinced others that he/she had "found the way,",who had good ideas, and who had both the substantive and political skills to the bridge the gap between the protagonists. Such persons have emerged in some important previous environmental negotiations, but not in the IWC. Perhaps someone outside the negotiations might be tapped to help break the stalemate.

Intervenor or Mediator

What might help break the deadlock is a person or persons who could act as an informal mediator, a facilitator, a point of contact; someone who all sides could agree has good ideas, is fair-minded, and is an authority figure; someone whose behavior is "reflective" and therefore can act confidentially, and is nonjudgmental. Unfortunately, if the IWC situation is not ripe for a negotiated agreement, it is also not ripe for formal, binding mediation or arbitration. Establishing the worth of a mediatory figure would probably be slow and painful before it promises the possibility of success.

Finding a mediatory leader or leaders outside of a negotiation is not mandatory, but it will be very difficult finding one within. A cornerstone of Canadian and Norwegian foreign policies has been determination to play the role of intermediary during the Cold War between the superpowers, and between the United States and the Third World in many international organizations. Alas, while trusted to be fair-minded in these other situations, they would not be so regarded in the IWC.

But there are rich possibilities for transforming a "dyad" into a triangle — that is, add a third party to the two antagonists, someone(s) who can help restore communications, formulate new ideas for resolution, and even manipulate the situation. Individuals of international prestige might be asked to act as facilitator, perhaps Tommy Koh, president of both UNCLOS III and the Rio Earth Summit, or Maurice Strong (though a Canadian), or Elliott Richardson of the United States. There is no one more trusted than Dr. Arvid Pardo, the "Father of the Law of the Sea negotiations," and winner of the Third World Prize. If one wishes to be puckish, one could suggested Boutros Boutros-Ghali, now that he is unemployed. I'm sure there are others whose reputations for probity were earned outside the ocean or environment nexus who would also be suitable. Finally, if the parties at issue do not want to trust an individual to provide a path out of the thicket, it might be possible to form an "Eminent Persons Group" to suggest ways out.

Civility and Trust

It might seem contradictory to suggest that the pro-whalers become "tougher" while acknowledging that a major deterrent to resolution is existing incivility and distrust. Toughness often exacerbates human feelings and could make personal relationships worse, but relations are so bad now that I'm not sure they could deteriorate further. But, it is possible to be both firm and fair-minded; to have a reputation for "good faith." It is important to have a reputation for a willingness to listen to others.

Too often interactions are merely occasions to restate one's own position. Too often the parties to the quarrels in the IWC sponsor meetings or seminars that preach to the converted. It is now time to ask oneself: what in others' positions is a legitimate demand to which I could accommodate myself? What do we have in common? Can we hold a discussion which starts from common ground and works outward until we reach that border area where discussions slip from agreement to disagreement so that we can narrow the range of disagreement? After all, no one in the current whaling debate begins from the position that unrestricted whaling should be restored. Perhaps a series of unofficial seminars, symposia, or workshops could bring together individuals representing both governments and NGOs to explore common positions or develop new common positions. It should be possible to borrow from environmental mediation efforts or arms control confidence rebuilding techniques methods for bringing the parties closer together.

To be blunt, the anti-whalers are in the saddle, therefore the pro-whalers must do more than try to meet them halfway. Anti-whaler governments and NGOs have no reason to change their position other than fearing that if they press too hard the whole system might collapse. They know they are in a superior bargaining position. I suggest they have some legitimate concerns that must be assuaged. The problem must be approached from the perspective of what can be done to foster the breakup of the anti-whaling coalition. Not all anti-whalers are amenable to a reasoned interaction on the issues; those whose position is primarily normative are likely to remain obdurate. Those whose concerns about whales and whaling are based on utilitarian considerations or the historical record might be willing to listen. It might be insulting to be regarded as unreformed predators but pro-whalers must recognize that they will never get whaling in any form reauthorized internationally until they provide ironclad assurances that whaling will never again get out of control. Such assurances cannot come from the pro-whalers without consultation with the anti-whalers.

Although I think the burden of proof in demonstrating the sustainability of whaling rests on the shoulders of whaling proponents, they have legitimate concerns to which the anti-whalers must also be responsive. Anti-whalers should try to understand the impact of a rigid anti-whaling position on smallholders, indigenous and non-indigenous alike, who are attempting to work out a painful compromise between sustaining their cultures and lifestyles and living in a world system of urbanized values; of the arrogance of assuming that one set of "moral" requirements fits all. Civility is a two-way street.

Complete the RMP and RMS

In recent IWC meetings, the major pro-whaling states have played "after-you Alphonse-Gaston", on completing the Revised Management Procedure and the Revised Management Scheme — each waiting for the other to introduce something new leading toward completion of the necessary formal revisions to be entered into the Schedule. While the Procedure is largely complete, there may be refinements not considered part of the Scheme that could be introduced. They should be evaluated, and if they are a measurable improvement, submitted. While some pro-whalers believe that the Scheme is "overkill," they must submit some of their own ideas to complete the Scheme. Not doing so will lead to a "Catch-22." Anti-whaling forces will submit nothing and claim that pro-whalers also submitted nothing and are not cooperating, so it is appropriate to not approve whaling. Or the anti-whalers will submit proposals that are so unrealistic and expensive that the pro-whalers will have no choice but to reject them.

Quiet Diplomacy

Since it is so difficult to resolve the fundamental splits on the floor of the IWC's annual meetings, perhaps the concerned governments should expend more effort at finding a resolution off the floor. Open covenants, openly arrived at often means no covenants arrived at.

Some types of interactions already take place off the floor, but they are used mainly to firm up positions on specific issues already on the IWC agenda. These are visits by Commissioners or delegation members to the national capitals of other states, including allied and opposing states. Coalition meetings occur. What is needed is not just intersessional firming up of positions, but an opportunity to get the major protagonists together to rethink the overall fate of the IWC on an informal basis, to mull over what can be done to assure the future of the IWC and the viability of a universal regime for whales. When matters have reached an impasse, it is time to think big.

These remarks may trigger in some readers the reaction: here goes another attempt to write a new constitution for the IWC, a new treaty. This type of effort failed before. Would it fail again? Perhaps, but the situation is close to 20 years further down the road, it is even more gridlocked, and unless someone blinks, it won't change. With each passing year, there is a greater chance of defection. A well planned informal session of senior officials away from the direct pressure of NGOs (althought their ideas must be solicited) need not result in a new treaty. There are ways that fundamental changes can be made other than a formal rewriting, resigning, and reratification of a convention.

There are hints that some governments are thinking along these lines, a rumor that some governments would be interested in hosting/participating in an informal session that could announce major changes in the IWC to coincide with the 50th anniversary of the Commission's founding. The changes would not require a formal renegotiation of the treaty. There are also rumors that some pro-whaling governments have dismissed such a notion, fearing that it is a device to strengthen the IWC mandate. Perhaps such a meeting would be a trap. But the participants would be sovereign states. If they did not agree, they could walk out. I don't think that any signal that some of the major players might want to try to untangle the standoff should be dismissed out of hand.

Rethinking National Position

Creative new ideas are unlikely to emerge from the business-as-usual way of thinking that has dominated the current interaction. Often there are alternative ways of achieving an outcome that protects fundamental interests if the focus is trying to understand the interest, rather than worrying about whether varying one´s present position ever so slightly would be perceived as retreat. To be most effective, those making suggestions must ask not only "what do our sponsors care about, but what do others care about." Can they invent options that would provide for mutual gains? This could be done informally by, as appropriate to each national situation, appointing agency or interagency study groups within governments with a mandate of providing a fresh look, commissioning outsider studies to seek new concepts, new language, asking intergovernmental organizations with expert knowledge but no direct interest at stake. It might help if some of these groups include not only "neutrals" but also individuals from committed groups with a personal reputation for fair-mindedness and a willingness to listen and contribute. Again, there are many possible permutations, as long as the stakeholders do not, as the police chief in the movie *Casablanca* orders, "round up the usual suspects."

If by some stroke of good sense stakeholders are willing to look to the premises that underlie their positions and see that there are other ways of expressing their interests, it also might be possible to adopt new negotiating tactics that could contribute importantly to a positive sum outcome. For many years Roger Fisher and associates of the Harvard Negotiating Project proposed a method called "principled negotiation." Many of the ideas expressed above are drawn from their publications: separate people from the problem, focus on interests, not positions, and invent options for mutual gain. They claim that their methods are neither hard nor soft but oriented toward finding efficient ways of reaching consensus on mutual gains. Below, I will outline some "hard" positions. If Fisher is correct, his bargaining methods can still be useful even if it becomes necessary to follow "tougher" tactics. In any case, all parties would benefit if they got away from defending positions and if in so doing, they found a jointly acceptable outcome.

Distributional vs. Integrative Negotiations

If the delegates and NGO representatives to IWC were to develop a set of principled strategies, it is likely that the IWC negotiation would move from a zero-sum distributive to positive-sum integrative outcome. There would be more to share, although there will always be some distributive elements, even in an integrative outcome. I bring the discussion back to the distributive-integrative distinction for several reasons. First, even if an outcome is largely integrative, there will always be distributive elements. In practice this could mean that, under the best possible alternative, the minority may have to give up more, and the majority less to come to a positive-sum outcome. Second, the conversion does not happen by itself. Good tactics will be necessary. I. William Zartman notes that there are only three known ways to foster the conversion: expanding the common pie, establishing trade-offs between differently-valued articles, or producing side-payments. These tactics can succeed only if parties concede that the "other" has legitimate interests and concerns. Finally, I return to distributive-integrative because, if by a stroke of luck or genius, it is possible to make the transition, all parties must recognize that a transition is taking place. The opportunity should not be wasted.

Act Strategically

Until it is clear that the anti-whalers are interested in an integrative outcome, pro-whaling states and coalition should act strategically: use tactics most likely to lead to the goal, even if they require that pro-whalers violate some of their substantive values. To this point, the anti-whalers have counted on the "sincere" behavior of the pro-whalers; that is, they expect that pro-whalers will make and execute their decisions based on the merit of the proposal. Strategic behavior calls for judging the issue on whether a positive vote will further *your* goals regardless of merit. So, if the United States proposes a quota for its Alaskan native whalers, strategically your vote would depend upon what promises/actions the U.S. makes in relation to issues highly salient to you. It need not even be directly tied to a specific issue. To make their point about their interests being ignored, a dissident group can vote no on everything, substantive or procedural. They can tie up the meeting if they are clever in manipulating the rules of procedure. They can refuse to or delay paying their dues until a new formula is developed that puts the IWC support burden on anti-whaling states, since the only income under the conditions of no whaling is from whale watching. Or propose a international tax on whale watching, etc. Again, hopefully, it need not come to Commission breakdown, but demonstrating that minority needs must be considered could have a salutary effect.

Strategic behavior also calls for treating the IWC meetings as bargaining, not problem- solving sessions. Bargaining is a combination of offers and threats. There have been precious few offers or threats made by the pro-whalers.

What kind of threat could be made that would be meaningful to anti-whalers? Obviously, the threat of pro-whaler withdrawing from the IWC is always there. Unrestricted whaling is not a credible threat, but the creation of a rival organization might be. The North Atlantic Marine Mammal Commission (NAMMCO) was created partially as a threat, but it has been held on a very tight leash. It is possible to loosen the leash. Again, creative moves could potentially raise the cost of gaining all anti-whalers' goals.

IWC politics are based on coalition. As noted, the anti-whaling coalition is very stable. If the pro-whalers are to get anti-whaling states to listen to their arguments, key members must see the pro-whaler coalition as a bit shaky and subject to possible defections. It is difficult and costly to break up a successful coalition, but it has been done. In the case of the IWC, it might be possible to "repack" the membership, that is, bring in enough new states favorable to one's position to form a new majority. There are many more potential whaling "stakeholders." A 1977 IWC study showed that 102 countries had whale stocks off their coasts. Since their interests are not being represented, it would be "democratic" as well as self-interested to solicit their participation. Many are Middle Developing States (formerly called Third World) who have a strong interest in protecting their right to exploit their natural resources. While whale resources are "common property" in areas beyond national jurisdiction, the Middle Developed States, and their U.N. caucusing group — the Group of 77 — might view the push toward imposing western urbanized values re whaling as a bad precedent and potential threat to their interests. This would be a very large Genii if let out of the bottle.

Promoting/Relying upon External or Exogenous Change

Lobbying by representatives of one government before the administrative agencies, legislature, or public of another government was considered by traditional diplomats to be a grave violation of their proper role as diplomats. Monsieur De Callieres would never approve, but it is much done in our own age. The object of such efforts is to put pressure on the government under assault to take measures favorable to the lobbying government, or if it has not, to change its position. Promoting change in another government's policy is common today, but convincing another government to reverse a set position is quite difficult, costly, and risky. It should be approached with caution.

This leaves the waiting game — relying upon the weight of time to induce position change because the problem has changed, the personnel have changed, etc. Relying upon a process over which one does not control is accepting the weakness of one's position. This seems to me to be the present situation. I doubt if internal change due to changing values will make any difference in the preferences of anti-whaling state officials or public in the foreseeable future. But just hanging on and hoping to wait out one's opponents appears to be a low cost option. Thus far, it has not worked, and I doubt if it will.

Relying on the United States

The United States should not be relied upon to salvage the situation, much less provide entrepreneurial leadership, in future whaling negotiations. In discussions with pro-whaling leaders, I detect both a lingering hope that it might, and exasperation that the U.S. so often takes contradictory positions. The hopeful point to the U.S. as the remaining hegemon, with a stake in assuring a positive outcome for whaling, and also that whaling does not put further pressure on the fragile political/economic alliance structures of the Cold War that the U.S. would now like to transform to fit the needs of a post-Cold War world. In this environment, the U.S. ability to command the loyalty of others is shakier than in the Cold War past. Conversely, so is the ability of its allies to put pressure on the U.S. to look favorably on *their* interests.

The main reason not to rely on the United States is because it is seriously cross-pressured. It has within its voting public large lobbying organizations with a commitment to end whaling and a community and state delegation (Alaska) that insists that the United States accommodate their needs. Since an exception to the no-whaling rule is available under the Whaling Convention, the United States uses it. But some raise the question how the U.S. can in good conscience lead the effort to end whaling and at the same time preserve whaling for some of its citizens privileged by an exception already written into the Schedule? Or expand that to other of its citizens — the Makah Tribe of Washington state — who have treaty-based whaling rights?

Unfortunately the United States has a long record of establishing a principle and then either seeking exceptions from the principle for itself and its citizens or ignoring the substance of the principle when it does not suit it. But advising not to rely upon the U.S. to salvage the negotiations does not mean the U.S. can be ignored. Perhaps because it must protect the rights of some of its citizens to continue to take whales, it can be brought to realize that others have needs as well.

Preparing a BATNA

Most negotiation theorists emphasize the need for preparing a BATNA or Best Alternative To a Negotiated Agreement. As Fisher and Ury note: "If you have not thought carefully about what you will do if you fail to reach an agreement, you are negotiating with your eyes closed." [1] A negotiator must know whether what is on the table will make their principals better or worse off. If worse off, they must look to other options. Moreover, it is important to convey to those on the other side of the table that you have options. Such recognition may help to soften an opponent's position. While the development of NAMMCO was motivated partly to demonstrate that its members have a BATNA, more must be done to show anti-whaling forces how they might lose control of the situation unless they are willing to negotiate seriously. Indeed, exercising some of the preliminary steps (hopefully reversible if appropriate) of a BATNA may be called for, such as having NAMMCO "authorize" controlled whaling. This might be combined with publicly available plans showing how NAMMCO could be the operating arm of a regional organization whose activities can be coordinated with the IWC. In any case, more detailed planning on BATNA options is now appropriate both for its own sake and as a bargaining tool.

CONCLUSION

I wish I could promise that, if followed carefully, this advice to the Princes would guarantee a successful conclusion to the effort to manage whales and whaling into the 21st century. Alas, I can only hope it has been helpful. At least, considering one or more measures recommended could not make matters worse.